Biostatistics

Concepts and applications
for biologists

For Tom

Biostatistics

Concepts and applications
for biologists

Brian Williams

London School of Hygiene and Tropical Medicine, UK

CRC Press
Taylor & Francis Group
Boca Raton London New York

CRC Press is an imprint of the
Taylor & Francis Group, an **informa** business

A CHAPMAN & HALL BOOK

First published 1993 by CRC Press
Taylor & Francis Group
6000 Broken Sound Parkway NW, Suite
300 Boca Raton, FL 33487-2742

Reissued 2018 by CRC Press

© 1993 by Taylor & Francis
CRC Press is an imprint of Taylor & Francis Group, an Informa business

No claim to original U.S. Government works

A Library of Congress record exists under LC control number: 93003336

Publisher's Note
The publisher has gone to great lengths to ensure the quality of this reprint but points out that some imperfections in the original copies may be apparent.

Disclaimer
The publisher has made every effort to trace copyright holders and welcomes correspondence from those they have been unable to contact.

ISBN 13: 978-1-138-10490-7 (hbk)
ISBN 13: 978-1-138-55777-2 (pbk)
ISBN 13: 978-1-315-15031-4 (ebk)

Visit the Taylor & Francis Web site at http://www.taylorandfrancis.com and the CRC Press Web site at http://www.crcpress.com

Contents

Preface

This book is an introduction to statistics for students of the life sciences, particularly for field and laboratory biologists, but also for those studying agriculture and medicine. Students commonly face the statistical analysis of their data with anxiety, and yet statistics can be elegant and exciting. Statistical analysis can reveal subtle patterns in otherwise uninformative data, reduce large data sets to a few key parameters, give researchers confidence in their conclusions and ensure that they do not deceive themselves when interpreting their data.

This book stresses the concepts and ideas that underlie the most important statistical methods used in biology. I have placed these concepts and ideas in a narrative structure and consider the same problems from different angles and in progressively more detail. Once you understand the general principles and are able to apply them to specific problems, you should be able to analyse new problems and to discover new applications for statistics in your own work. To ensure that the text remains firmly rooted in real problems, I have taken examples and illustrations from the biological literature wherever possible.

When using statistics to help answer a biological question, it is important to understand the question clearly. Once you know what it is that you wish to gain from your data, statistical techniques become powerful tools for revealing information that is otherwise hidden. Always remember that the statistics is simply a way to help you think clearly about the biology.

FURTHER READING

Many books on biostatistics are available; some can be used to complement this book. I have drawn on the following sources.

Introductions

Bailey N. T. (1974) *Statistical Methods in Biology*, English Universities Press, London.
Glantz, S. A. (1987) *Primer of Biostatistics*, McGraw-Hill, New York.
Parker R. E. (1986) *Introductory Statistics for Biologists*, Edward Arnold, London.
Phillips, J. L. Jr. (1982) *Statistical Thinking*, Freeman, New York.
Porkess, R. (1988) *Dictionary of Statistics*, Collins, London.

Intermediate

Bliss, C. I. (1967) *Statistics in Biology*—Volume I, McGraw-Hill, New York.

Bulmer, M. G. (1979) *Principles of Statistics*, Dover, New York.

Campbell, R. C. (1987) *Statistics for Biologists*, Cambridge University Press, Cambridge.

Fleiss, J. L. (1986) *The Design and Analysis of Clinical Experiments*, John Wiley, New York.

Freund, J. E. (1972) *Mathematical Statistics*, Prentice-Hall, London.

Hays, W. L. (1988) *Statistics*, Holt, Rinehart and Winston, Orlando.

Jeffers, J. N. R. (1978) *An Introduction to Systems Analysis: With Ecological Applications*, Edward Arnold, London.

Manly, B. (1986) *Multivariate Statistical Methods: A Primer*, Chapman & Hall, London.

Mead, R. (1988) *The Design of Experiments: Statistical Principles for Practical Applications*, Cambridge University Press, Cambridge.

Mead, R. and Curnow, R. N. (1983) *Statistical Methods in Agriculture and Experimental Biology*, Chapman & Hall, London.

Siegel, S. and Castellan Jr. N. J. (1988) *Nonparametric Statistics for the Behavioral Sciences*, McGraw-Hill, New York.

Sokal, R. R. and Rohlf F. J. (1981) *Biometry*, Freeman, New York.

Sokal, R. R. and Rohlf F. J. (1987) *Introduction to Biostatistics*, Freeman, New York.

Sprent P. (1990) *Applied Nonparametric Statistical Methods*, Chapman & Hall, London.

Zar J. H. (1984) *Biostatistical Analysis*, Prentice-Hall, Englewood Cliffs.

Advanced

Green P. E. (1978) *Analyzing Multivariate Data*, Dryden Press, Hinsdale, Illinois.

Mason, R. L. Gunst, R. F. and Hess, J. L. (1989) *Statistical Design and Analysis of Experiments: With Applications to Engineering and Science*, John Wiley, New York.

Seber, G. A. F. (1989) *Linear Regression Analysis*, John Wiley, New York.

Snedecor, G. W. and Cochran, W. G. (1980) *Statistical Methods*, Iowa State University Press, Ames, Iowa.

Winer B. J. (1971) *Statistical Principles of Experimental Design*, McGraw-Hill, New York.

STATISTICAL SOFTWARE

Many statistical packages are available for personal computers and spreadsheet packages offer basic statistical routines. Using one of these you will avoid having to carry out detailed and tedious calculations. It is essential to have access to at least one such package to use in conjunction with this book. This is particularly important for the last two chapters on the analysis of variance and regression.

There are two very large packages called SAS and SPSS. These will do everything you ever thought of and a lot more besides. There are also a range of medium-sized packages that should be adequate for your needs and several small packages that are cheap but limited. Some vendors offer cut-down versions at cheap rates for students. Prices for the big packages range up to a thousand pounds and you may have to pay an annual licence fee; for smaller packages you should expect to pay up to several hundred pounds.

It is difficult to make firm recommendations because statistical software packages are developing rapidly. In this age of microcomputers, I believe that menu-driven software combined with publication quality, interactive graphics and a reasonable range of standard analytical procedures is essential. If you use a PostScript printer, you should check the printer drivers available in each package – even now some software does not support PostScript. Based on my own experience I would advise you to consider, at the very least, MiniTab, StatGraphics, StatView and SyStat. You should review all the packages carefully and make a decision based on your needs.

BASS from Bass Institute Inc., P.O. Box Chapel Hill, NC 27514, USA. Telephone (919) 489 0729.

C-Stat from Cherwell Scientific Publishing, 27 Park End Street, Oxford OX1 1HU, UK. Telephone (0865) 794884/794664.

Crunch Statistical Package from Crunch Software, 5335 College Avenue, Suite 27, Oakland, CA 94618, USA. Telephone (415) 420 8660.

CSS from StatSoft, 2325 East 13th Street, Tulsa, OK 74104, USA. Telephone (918) 583 4149.

Data Desk from Odesta Corp., 4084 Commercial Avenue, Northbrook, IL 60062, USA. Telephone (800) 323 5423.

Exstatix from Strategic Mapping, 4030 Moorpark Avenue, Suite 250, San Jose, CA 95117, USA. Telephone (408) 985 7400.

GLIM from NAG Ltd., Wilkinson House, Jordan Hill Road, Oxford, UK. Telephone (0865) 511245. Fax (0865) 310139.

Mac SS/Statistica from StatSoft, 2325 E. 13th Street, Tulsa, OK 74104, USA. Telephone (918) 583 4149.

MiniTab Statistical Software from Minitab Inc., 3081 Enterprise Drive, State College, PA 16801, USA. Telephone (800) 448 3555 and CLE Com Ltd., The Research Park, Vincent Drive, Edgbaston, Birmingham B15 2SQ, UK. Telephone (021) 471 4199.

NWA Statpak from Northwest Analytical Inc., 520 N. W. Davis Street, Portland, OR 97209, USA. Telephone (503) 224 7727.

S-Plus from Statistical Sciences Inc., 1700 Westlake Avenue N., Suite 500, Seattle, WA 98109, USA. Telephone (206) 283 8802. Fax (206) 283 8691 and Statistical Sciences UK Ltd., 52 Sandfield Road, Oxford OX3 7RJ, UK. Telephone (0865) 61000. Fax (0865) 61000.

SAS and *JMP* from SAS Institute, SAS Circle, Box 8000, Cary, NC 27512, USA. Telephone (919) 467-8000 and SAS Software Ltd., Wittington House, Henley Road, Marlow SL7 2EB, UK. Telephone (0628) 486933.

Solo 101 and *BP-90* from BMDP Statistical Software Inc., 1424 Sepulveda Boulevard, Suite 316, Los Angeles, CA 90025, USA. Telephone (213) 479 7799.

SPSS/PC+ from SPSS Inc., 444 N. Michigan Avenue, Chicago, IL 60611, USA. Telephone (312) 329 3300, and SPSS International, P.O. Box 115, 4200 AC,

Gorinchem, The Netherlands. Telephone (31) 1830 367 11. Fax (31) 1830 358 39.

StatGraphics from STGC Inc., 2115 E. Jefferson Street, Rockville, MD 20852, USA. Telephone (301) 984 5000, (301) 592 0050.

Statistix II from NH Analytical Software, 1958 Eldridge Avenue, Roseville, MN 55113, USA. Telephone (612) 631 2852.

StatPac Gold from StatPac Inc., 3814 Lyndale Avenue S., Minneapolis, MN 55409, USA. Telephone (612) 822 8252 and Perifernalia, Snoekstraat 69, Alken, B-3570, Belgium. Telephone (32) 11 313754.

Statview II/Super ANOVA and *Statview SE and Graphics* from Abacus Concepts, 1984 Bonita Avenue, Berkeley, CA 94704, USA. Telephone (800) 666 7828; (415) 540 1949.

SyStat/SyGraph from SyStat Inc., 1800 Sherman Avenue, Evanston, IL 60201, USA. Telephone (708) 864 5670.

Acknowledgements

I thank Catherine Campbell, Helena Cronin, Philip Loder, Susan MacMillan, Sarah Randolph, Daniel Remenyi and David Rogers for their comments, advice, support and encouragement. The quality of the text has been greatly improved by their labours. I am enduringly grateful to Russell Cheng, who taught me much of what I know about statistics. I thank MiniTab Statistical Software for providing me with a copy of MiniTab under their author support scheme. I thank the Royal Society for a Guest Fellowship at the University of Oxford to work with David Rogers and Sarah Randolph, during which time much of this book was written. I thank Philip Loder for drawing Figure 8.5 and Jenny Mitchell for her patience in proof-reading the book.

1

Why mathematics?

It is interesting to contemplate a tangled bank, clothed with many plants of many kinds, with birds singing on the bushes, with various insects flitting about, and with worms crawling through the damp earth, and to reflect that these elaborately constructed forms, so different from each other, and dependent on each other in so complex a manner, have all been produced by laws acting around us. These laws, taken in the largest sense, being Growth and Reproduction; Inheritance which is almost implied by reproduction; Variability from the indirect and direct action of the conditions of life, and from use and disuse; a Ratio of Increase so high as to lead to a Struggle for Life, and as a consequence to Natural Selection, entailing Divergence of Character and the extinction of the less-improved forms.   C. Darwin (1906, p. 669).

Mathematics and biology are two of the oldest sciences, both dating back to the beginning of recorded history, but it is only in this century that they have come together in the creation of modern mathematical biology. Well before the development of our modern view of natural history, much effort had been devoted to the systematic classification of plants and animals; by the seventeenth century many species had been described and identified. In 1686 John Ray defined species as groups of similar individuals characterized by 'distinguishing features that perpetuate themselves in propagation from seed' and noted that 'One species never springs from the seed of another' (Burkhardt, 1981). In 1735 Linnaeus developed the two-name system of classification for plants and animals that is used to this day. Believing in the constancy of species, he said, 'There are as many species as the Creator produced different forms in the beginning' (Burkhardt, 1981). By the nineteenth century this view of species as immutable entities was beginning to weaken; although naturalists accepted that species were 'genuine entities in nature, constant in their essential characters', they now added 'subject to non-essential, accidental variation' (Burkhardt, 1981). Although the classification of species cannot be called mathematical, we cannot do anything until we have defined the entities we wish to study.

Having defined and classified species, it was natural to ponder the reasons why these particular species were found in the world. The idea of change within the otherwise fixed species was unavoidable in a world in which selective breeding of plants and animals was widely practised, and this led to intense debates concerning the origin of the natural world: was the world created as we see it now or was it created in a simpler form from which our present world has evolved? Nearly all religions postulate a process of creation and the idea of a progressive, albeit rather rapid, creation of the world is found in all cultures. But if the world has evolved from a 'simpler' state, do we need to invoke a 'higher power' to guide and direct the process of evolution or can we find a theory that allows us to explain evolution without recourse to external powers? If we can find laws of evolution, what kind of laws should they be? Should the laws of evolution be causal and mechanistic as are the laws of classical physics? The only physical law that depends explicitly on the direction of time is the second law of thermodynamics and this law predicts increasing disorder as time passes. How can we reconcile this with the apparent increase in order as time passes that we seem to see in biological evolution?

Two thousand years ago Aristotle reflected on the same problem (Ross, 1952): 'Why should not nature work, not for the sake of something, nor because it is better so, but just as the sky rains, not in order to make corn grow, but of necessity?...Why... should it not be...that our teeth come up of necessity—the front teeth sharp, fitted for tearing, the molars broad and useful for grinding down the food... [and] survived, being organized spontaneously in a fitting way; whereas those which grew otherwise perished and continue to perish?' Since Aristotle believed that chance would destroy rather than preserve the adaptations we find in members of a given population, he rejected his own proposal and continued: 'Yet it is impossible that this should be the true view. For teeth and all other natural things either invariably or normally come about in a given way; but of not one of the results of chance or spontaneity is this true'. In this discourse Aristotle is concerned with the problem of selection rather than evolution, but he touched on the idea that the best adapted organisms should survive while those less well adapted should perish. Two thousand years later we find Darwin wondering how undirected chance events could have produced the apparently directed change found in the fossil record.

Darwin's interest lay in the ways in which changes arose in natural populations and he knew that substantial changes can be brought about in animal and plant populations by cross-breeding and selection. But while the work of breeders, who systematically select desirable characteristics, hinted at a mechanism for the emergence of novelty, Darwin could not see how such apparently directed change could be brought about in nature, if we assume that she is neutral and does not regard any particular change as more desirable than any other.

The answer came to Darwin from a quite unexpected source, as he recorded in his autobiography: 'In October 1838,...fifteen months after I had begun my systematic inquiry, I happened to read for amusement Malthus on *Population*, and being well prepared to appreciate the struggle for existence which everywhere goes on, from long-continued observation of the habits of animals and plants, it at once struck me that under these circumstances favourable variations would tend to be preserved, and unfavourable ones to be destroyed. The result of this would be the formation of a new species. Here, then, I had at last got a theory by which to work...' (Darwin, 1958). Wallace (1905) took his inspiration from the same source: 'One day something brought to my recollection Malthus' *Principle of Population*...I thought of his clear exposition of 'the positive checks' to increase...which kept down the population....

It then occurred to me that these causes or their equivalents are continually acting in the case of animals also; and, as animals usually breed much more rapidly than does mankind, the destructions every year from these causes must be enormous in order to keep down the numbers of each species, since they evidently do not increase regularly from year to year, as otherwise the world would long ago have become densely crowded with those that breed most quickly.... Why do some die and some live? And the answer was clearly, that on the whole the best fitted live. From the effects of disease the most healthy escaped; from enemies the strongest, the swiftest, or the most cunning; from famine, the best hunters or those with the best digestion; and so on...... The more I thought over it the more I became convinced that I had at length found the long-sought-for law of nature that solved the problem of the origin of species.' (Cronin, 1991, provides a fascinating account of the development of Darwinian theory from Darwin and Wallace to the present day.)

Malthus (1970) realized that if the growth of a natural population were unchecked it would increase exponentially, i.e. the numbers would double in a fixed time and then double again and again in each equivalent time period. He also argued that our ability to increase food production would increase only arithmetically, i.e. we can at best increase food production only by the same absolute amount in any fixed time. Therefore, population growth would always tend to outstrip increases in food production. Since human populations do not always increase exponentially, Malthus sought to identify the factors that act to limit population growth. After examining the evidence from a number of countries, Malthus concluded that 'vice' (including abortion and infanticide) and 'misery' (including hunger and disease) acted as checks, although in later works he added 'moral restraint' (abstinence), which he hoped might one day replace 'vice' and 'misery' as a check to population growth. In modern terms, Malthus understood that the growth of all populations must eventually be limited by an increase in mortality or a decrease in fecundity with increasing density. Darwin and Wallace now took Malthus's essentially mathematical arguments and realized that for all biological organisms an exponential increase in the number of individuals would lead to a struggle for life in which many would die and few would survive. Given the natural variation between individuals in a given population, those that are in some sense better adapted to their environment than the others will be more likely to survive.

We can illustrate the consequences of exponential growth quite simply. We know, for example, that one tsetse fly weighs about 30 mg. We also know that the maximum rate of growth of a tsetse fly population amounts to a doubling in the numbers each month (making the tsetse fly a very slow grower by insect standards). So if the world contained only one pregnant tsetse fly, and if this initial population grew at the maximum possible rate, at the end of one year the numbers would have doubled 12 times, making a total of 2^{12}, or about 4000, tsetse flies. At the end of 10 years the population would have increased by this same factor of 4000 another ten times, making a total of 4000^{10}, or 10^{36}, tsetse flies, weighing about 3×10^{34} g, which is five million times the mass of the earth. Since tsetse flies have been around for rather more than 10 years, the number of flies in existence today is only a tiny fraction of the number that could have been produced. Clearly there must be a constant thinning of the population, and it seems reasonable to assume that if the individuals of a population vary, those individuals that have slightly advantageous characteristics will tend to survive and pass on their genes while those that possess characteristics that are slightly disadvantageous will tend to be eliminated along with their genes.

While evolution has resulted in the production of the most varied and wonderful organisms, just as remarkable is the extraordinary stability that we see in the wings of the dragon fly, for example, which have remained unchanged for millions of years. To account for this we argue that if a particular characteristic is optimal in the sense that any small change from that state is disadvantageous, there will be a tendency for natural selection to maintain that state. Darwin's theory of natural selection therefore attempts to explain both speciation and stasis.

Darwin's theory provided a convincing and detailed description of the processes by which biological evolution occurs and it removed the need for a teleological theory of evolution. Nevertheless, Darwin was aware that he did not have a **mechanism** that would explain the process of evolution at a deeper level. 'The laws governing inheritance are for the most part unknown', he wrote. 'No one can say why the same peculiarity in different individuals of the same species, or in different species, is sometimes inherited and sometimes not' (Darwin, 1906, p. 15). The key to the mechanism that Darwin lacked was discovered by Gregor Mendel, who in 1866 published the results of his work on inheritance in garden peas, *Pisum sativum* (Mendel, 1866). The modern theory that derives from Mendel's work is essentially 'atomistic'. Characteristics of what we now call the phenotype are conferred on organisms by a small number of inherited 'particles' that we now call genes. But although Mendel's papers were sent to the Royal and Linnaean Societies in England (Fisher, 1936), his work was almost entirely neglected, perhaps because his observations contrasted so strongly with the 'continuum' theory of pangenesis, proposed by Aristotle and accepted by Darwin (Ayala and Kiger, 1980) in which inheritance involves a blending of fluids created in the bodies of the parents. To put Darwin and Mendel's work into prespective, it is worth remembering that Schwann established the theory that living organisms were composed of separate cells only in 1839 and that cell division reached wide acceptance only after the work of Virchow in 1858 (Ronan, 1983). It was not until the middle of the nineteenth century that atomism achieved universal acceptance in physics and chemistry.

At the turn of the century Mendel's work was rediscovered, perhaps because the intellectual climate had become more favourable to his ideas. Biologists soon realized that the new science of genetics provided a quantitative basis for Darwin's essentially qualitative theory, leading to what we now call the neo-Darwinian synthesis. Genetics itself has been placed on a sound chemical basis during this century with the elucidation of the structure of DNA, our increasing understanding of the biochemical basis of genetic replication and the establishment of molecular biology and genetic engineering.

As biological thought has evolved over the past three centuries, biology has become more quantitative and precise. Although Darwin was able to develop his theory without mathematics, the essentially mathematical idea of exponential growth was crucial to the development of his ideas. Mendel was obliged to use statistics to analyse his experiments on peas, Watson and Crick needed advanced mathematics to determine the structure of DNA from Franklin's X-ray photographs and geneticists now rely on sophisticated mathematics in the analysis and interpretation of their experiments. Now we have come full circle: we are applying mathematics to organisms and their inter-actions, the basis of much of Darwin's work, in the modelling of ecosystems. Having started from Darwin's essentially qualitative theory of evolution, we have worked our way down to an understanding of the process of evolution at a molecular level. Now we are working our way back up again, synthesizing ideas and information from many different disciplines, at many different levels, and creating in the last few decades a quantitative theoretical ecology.

Over 300 years ago, Galileo Galilei revolutionized natural philosophy by systemati-cally applying mathematics to the analysis of experimental data. 'Philosophy', he said, 'is written in that great book which lies before our gaze—I mean the universe—but we cannot understand it if we do not first learn the language and grasp the symbols in which it is written. The book is written in the mathematical language,... without the help of which...one wanders in vain through a dark labyrinth' (Needham, 1972, p. 32). Biology is now following the same path and little can be done without mathe-matics. Despite the difficulties many of us have in learning mathematics, the power it gives us more than repays the effort it takes to learn it.

1.1 SOME PROBLEMS

Before launching fully into biostatistics, I shall introduce you to some of the problems discussed in this book. Here I hope to persuade you that the answers are worth knowing; in the chapters that follow we will address these problems and many others. By the time that you have finished reading the book, I hope that you will agree that a knowledge of statistics will help you to analyse otherwise intractable problems and to draw reliable and consistent conclusions from your experimental data.

1.1.1 The casino

Suppose that a friend of yours offers to play a game of chance with you. She will throw a pair of dice and for £1 you are to guess the sum of the two numbers that turn up. If your guess is wrong, she keeps the pound; if it is right, she pays you as many pounds as the number you guessed. What number should you guess? Should you play with her at all? If you make your best guess, will you win or lose, and if so, how much?

1.1.2 Mendel's peas

In his work on inheritance in peas, Mendel (1866) crossed a number of pure-bred tall plants (which have two 'tall' genes, $T T$) with pure-bred short plants (which have two 'short' genes, tt). The plants of the first filial generation, F_1, inherited a 'tall' gene from one parent and a 'short' gene from the other. Since the 'tall' gene is dominant all of the F_1 plants were tall. Mendel then crossed 1064 F_1 plants and in the second filial generation, F_2, he obtained 787 tall plants and 277 short plants, as shown in Table 1.1.

Table 1.1 The results of an experiment conducted by Mendel in which he compared the number of tall and short pea plants in the F_2 generation to the numbers he expected on the basis of his theory of inheritance. The expected numbers are $1064 \times 3/4 = 798$ tall plants and $1064 \times 1/4 = 266$ short plants

	Tall plants	Short plants	Total	Ratio tall/short
Observed	787	277	1064	2.84
Expected	798	266	1064	3.00

Now if the probability that either of the two F_1 parents will contribute a 'tall' gene to the F_2 generation is the same as the probability that either will contribute a 'short' gene, and given that the 'tall' gene is dominant, we can show that there should be three tall plants for every short plant, or 798 tall plants and 266 short plants in the F_2 generation. In section 2.2.1 we shall see why Mendel decided that there should be three tall plants for every short plant, but for the moment we note that the ratio of the number of tall to short plants that he observed was 2.84 rather than 3. Does this refute his theory that the ratio should be 3 or was it just chance that the observed ratio was not quite equal to the expected ratio? Suppose, for example, he had obtained 760 tall and 304 short plants, giving a ratio of 2.5. Would these data support or refute his theory? Indeed, by how much could the ratio deviate from 3 before we would have to say that the data do not support the theory?

1.1.3 The polio vaccine

In the early 1950s poliomyelitis was recognized as a major disease causing crippling paralysis and death, especially in children. In 1954 Jonas Salk developed a vaccine against polio (Snedecor and Cochran, 1989, p. 13). These days there is much talk about the possibility of developing an effective vaccine against malaria (Targett, 1991). If you were to carry out a trial of such a vaccine, how would you proceed?

In the study carried out to test the Salk vaccine, the results of which are given in Table 1.2, 200 745 children were vaccinated, of whom 33 developed polio. Those conducting the study needed to determine how many of these children would have developed the disease if they had not been given the vaccine. The vaccinated children

Table 1.2 The number of paralytic cases of polio that developed among two groups of children, one of which was given the Salk polio vaccine and the other of which was given a saline solution

Group	Number treated	Paralytic cases	Cases per 100 000
Vaccinated	200 745	33	16
Placebo	201 229	115	57

obviously could not be used to discover what would have happened if they had not been vaccinated, so a separate **control** group of children was injected with a saline solution placebo rather than the vaccine. Of 201 229 children given the placebo, 115 developed polio. The proportion of children who developed polio was greater among those given the placebo (57 per 100 000) than among those given the vaccine (16 per 100 000), so the vaccine appears to be effective in reducing the incidence of polio, although in both cases only a small number of children developed the disease. Any intervention, including vaccination, carries some risk. On the basis of these results, would you have your child vaccinated against polio? A trial involving nearly half a million children is both costly and time-consuming; could the same result have been obtained more quickly and at less cost by treating only 20 000 children in each category?

A number of other questions arise. How could the scientists carrying out the trial be sure that, as far as the vaccine and the disease were concerned, the control group of children did not differ significantly from the treated group of children? There is a moral dilemma: since the trial would not have been carried out unless the scientists already believed that the vaccine was effective in preventing polio, how should they decide which children to vaccinate and which to treat as the control? The design of experiments is crucial and forms an entire area of biomathematics in itself, although it is too often overlooked. Months of hard work may be devoted to doing an experiment in biology only to discover that because of a flaw in the design of the experiment or the presence of a factor that should have been included but wasn't, the data do not provide the answer to the question being asked. Indeed, the hallmark of a good scientist is the ability to design experiments that will elicit the subtle changes and effects one is seeking in research.

1.1.4 Controlling armyworm

The conventional way of controlling the African armyworm, *Spodoptera exempta*, is to spray crops with pyrethrum, but this pollutes the environment and kills many of the predators that feed on armyworms. A more attractive control strategy involves the use of biological control agents, such as the bacterium *Bacillus thuringiensis* var. *aizawai* (*B.t.*), which can be stored as a powder and applied in a water suspension.

In June 1988 an outbreak of the African armyworm was detected in a 40-acre wheat field at Ngungugu Farm, near Nakuru, in Kenya (Brownbridge, 1988). A suspension of *B.t.* in water was applied at concentrations of 0.5, 1.0 and 2.0% weight by volume, to three different parts of a field infested with armyworm, with the results shown in Table 1.3. *B.t.* was not applied in the control area. In the control area there is no obvious reduction in the number of armyworm larvae over 4 days. Treatments *A* and *B* both appear to reduce larval numbers by a factor of three after 4 days, while treatment *C* reduces larval numbers by a factor of about 75 after 4 days and appears to be very effective. But are the slight reductions under treatments *A* and *B* significant or are they simply due to chance? Do treatments *A* and *B* differ from each other? For how long would we have to apply each of the treatments to achieve a 99% reduction in the number of larvae?

Table 1.3 Mean counts of armyworm larvae on each day after applying three different treatments: Control—no treatment, A—0.5%, B—1.0% and C—2.0% *B.t.* suspended in water

| | | Treatment | | |
Time/days	Control	A	B	C
0	497	320	294	295
1	463	203	213	93
2	506	155	118	33
3	487	125	124	12
4	480	101	111	4

1.1.5 Summary

When you have finished reading this book, I hope you will be able to answer all of these questions without difficulty. These problems were chosen because they are easy to state and to appreciate but we will encounter more complicated problems. For example, we have only considered experiments in which the outcome is determined by a single factor: whether or not the children were vaccinated, whether the peas had tall or short genes. But more often than not we will be analysing data in which there are many factors that may determine the outcome of an experiment, and we will want to separate out the contributions of the various factors as best we can. Whenever possible, we will design our experiments in such a way that the contributions of the various factors that determine the outcome are independent of one another so that we can assess their effects independently.

1.2 A BRIEF OUTLINE

In this book you will encounter new words—Poisson distributions, cumulative distribution functions, analysis of variance, regression, and many others. Just as we use Latin names to indicate particular species and thus avoid lengthy taxonomic descriptions, we need to do the same in biostatistics so that we can express sophisticated ideas clearly and concisely. This means thay you will have to learn a new language of words and symbols. We will, for example, define probability rather carefully, so that when we say, 'If I spin a coin, the probability that it falls heads is one half', we know precisely what this means. We will then abbreviate the statement to: $P(H) = 0.5$.

Often the variables we are studying are beyond our control. For example, the number of tsetse flies caught in traps is high during the rains (Dransfield *et al.*, 1990). But the humidity is also high during the rains, so is it the rain or the humidity causing the increase in the catch? In fact, when it rains the temperature is generally low and perhaps the high catch has nothing to do with rain but rather is related to low temperatures. Much of biostatistical theory is devoted to developing techniques for separating out the effects of different factors, in this case rainfall, temperature and humidity, on the variable of interest, in this case the number of tsetse flies caught in a trap, the variation of which we hope to understand.

The book is divided into the following main sections.

- Probability—the fundamental concepts and ideas underlying the application of statistics to biology.
- Representations—ways of picturing distributions of random numbers.
- Measures—ways to measure and describe distributions of random numbers.
- Basic distributions—the most important distributions of random numbers.
- Testing hypotheses—how to frame questions and test predictions.
- Comparisons—how to make reliable comparisons among the results of experiments.
- Analysis of variance—how to analyse variations in data in terms of contributions from the separate factors that may affect and influence the data.
- Regression—how changes in one variable are related to changes in another.

It is unfortunate, but true, that many biologists find statistics difficult and there is no doubt that the statistical analysis of biological experiments can become very complicated. However, the number of concepts underpinning biostatistics are few.

- There are **two fundamental laws** of probability and **two laws for combining probabilities**.
- We need to understand the concept of **probability distributions** and their measures. The two most important measures are the **mean** and the **variance**. The mean is the average value; the variance is a measure of the 'spread' of the observed values about the mean.
- When several factors may affect the result of an experiment, we want to know how much of the variation in the result can be ascribed to each factor. For example, if we feed male and female rats on fresh and rancid lard, we might want to know how the sex of the rats as well as the freshness of the lard affects the weight gained by the rats.
- When we have several things going on at once, for example, crop growth, fruit production, rainfall and insect infestation, each of which may be important, we want to know how they are related, and which ones influence the others. We might say, for example, that rainfall and crop growth tend to increase or decrease together, while insect infestation and fruit production are inversely related, so that as the one increases the other decreases.
- We want to be able to measure variables, such as rainfall and temperature, and use these to predict other variables, such as the growth of a crop. But we want to be able to pick out those factors that affect the outcome, in this case the growth of a crop, most strongly. If we then decide that the key factors are rainfall and temperature, we want to find the combination of the two that gives us the 'best' prediction of crop growth, and we need to think about what we mean by the 'best' prediction.

These few ideas really contain everything that is covered in this book. If you can get a firm grasp of the key underlying ideas, then the subtle and complex applications should not be difficult. If you lose sight of the underlying generalities, however, even relatively simple applications will seem confusing and difficult. On the other hand, it is also important to develop technical skills because you do not always want to start from the laws of probability. The English philosopher and mathematician Alfred Whitehead (1928) put it like this: 'Without a doubt, technical facility is a first requisite for valuable mental activity: we shall fail to appreciate the rhythm of Milton, or the passion of Shelley, so long as we find it necessary to spell the words and are not quite certain of the forms of the individual letters. In this sense there is no royal road to learning. But it is equally an error to confine attention to technical processes, excluding consideration of general ideas. Here lies the road to pedantry.'

Taylor and Wheeler (1963) gave the following advice to young physicists: 'Never make a calculation until you know the answer. Make an estimate before every calculation, try a simple physical argument... before using every derivation, guess the answer to every puzzle. Courage: no one else needs to know what the guess is. Therefore, make it quickly, by instinct. A right guess reinforces this instinct. A wrong guess brings the refreshment of surprise.' I hope that you too will have the courage and the confidence to guess.

1.3 EXERCISES

1. In an experiment growing duckweed under controlled conditions in the laboratory, it was found that the number of fronds doubles every 2 days. Starting with one frond, how many would there be after 1, 2 and 3 months if the growth continued unchecked?

2. The formation of locust swarms tends to occur after drought-breaking rains. The process of swarm development involves increased rates of multiplication, concentra-

Table 1.4 The results of a series of hypothetical experiments carried out to test Mendel's hypothesis that the ratio of tall to short plants in the F_2 generation is 3

Number of tall plants	Number of short plants	Ratio
30	10	3.0
28	12	2.3
26	14	1.9
24	16	1.5
22	18	1.2
20	20	1.0
300	100	3.0
280	120	2.3
260	140	1.9
240	160	1.5
220	180	1.2
200	200	1.0

tion of the locusts into small areas and behavioural and morphological changes from the 'solitarious' to the 'gregarious' phase. In a study of the origin of plagues of the South American locust *Schistocerca cancellata* in Argentina, Hunter and Cosenzo (1990) showed that plagues began with a season when three generations per year were possible and plagues were sustained when only two generations were possible. Plagues typically took two years to increase to maximum size. The number of generations per year is therefore an important factor in determining the outbreak and decline of plagues. Assuming that each female locust lays about 80 eggs in a pod of which about 30 survive to adulthood (Farrow, 1979), calculate the factor by which the number of locusts could increase in two breeding seasons with one, two and three generations per breeding season.

3. Imagine that you repeated Mendel's experiment (section 1.1.2). Guess which of the (hypothetical) results given in Table 1.4 support his hypothesis that the expected ratio of tall to short plants in the F_2 generation is 3 to 1? (You will find the answers in Chapter 6, Exercise 1.)

4. Plot graphs of the number of armyworms and of the logarithm of the number of armyworms against time from the data in Table 1.3 for the control and the three treatments. What do you notice about the two sets of plots?

2

Probability

... young students, familiar with animals, ... are often disappointed that their painstaking
analysis, above all the dull statistical evaluation, finally shows nothing more than what
a sensible person with eyes in his head and a good knowledge of animals knows already.
There is, however, a difference between seeing and proving, and it is this difference
which divides art from science.   K. Lorenz (1967)

A curious feature of science is that the foundations, which most people assume are
well established, are the least secure parts of the edifice. When we begin a new study,
we generally make a few vague statements to get us going and then race happily
onwards. The theory of probability and statistics is no exception and the philosophical
foundation of the theory is the subject of intense debate. A reasonably thorough
discussion of the foundations of statistics would require an entire course in itself, so
I shall simply give you my own biased view of how best to get going. (Hacking,
1979, discusses the foundations of statistics in his book *Logic of Statistical Inference*.)

2.1 THE CONCEPT OF PROBABILITY

Suppose that I spin a coin and ask, 'What is the probability of getting heads rather
than tails'? Most people would say 50% or one-in-two. But suppose we try to be more
precise and ask ourselves what we mean by this. There are several ways in which
we could expand the answer further.

One way would be to say that we are just as likely to get heads as to get tails.
But this only says the same thing in different words and what we probably mean is
that we simply have to take the definition as given, that we cannot explain it further.

Another explanation might be to say that if we throw a coin very many times,
we will get heads on about half the throws. To test this, I threw a penny one million
times (I confess with the aid of a computer pretending to be a penny) and had
500 737 heads and 499 263 tails. The proportion of heads is 50.07%, which is almost,
but not exactly, equal to the expected 50%. So how close to 50% do we have to be
before we can say that heads and tails are indeed equally likely?

Still another explanation might be that if someone throws a coin, it is worth betting
that it will land heads only if the odds you are offered are better than 1 to 1.

These answers represent three ways of looking at probability. The first, *a priori*,
approach says that the concept cannot be reduced further and so we simply have to

accept that a thing called probability exists and then proceed to define laws for combining probabilities, derive theorems about probabilities (much as you may have done
with Euclidean geometry at school) and then try to match our theoretical predictions
with the outcome of whatever experiments we do.

The second, empirical or experimental, approach says that we should define probability in terms of **long-run frequencies**. This is certainly a useful way to think about
probability and is the way most people do think about it. In the first edition of his
book *Logic*, the English philosopher John Mill wrote (Bulmer, 1979, p. 5), 'Why in
tossing up a halfpenny do we reckon it equally probable that we shall throw heads
or tails? Because we know that in any great number of throws, heads and tails are
thrown equally often; and that the more throws we make, the more nearly the equality
is perfect.' But this long-run approach is not without its problems. We might agree
that there is a definite probability that I shall be run over by a bus tomorrow: but
if it does happen, I shall hardly be in a position to repeat the experiment many times.

The third way of looking at probability, which we might call the casino definition,
is particularly suited to those who gamble. While I would not like to suggest that
Mill led a dissipated life, he explicitly rejected the frequency definition in the later
editions of his book and gave the following definition instead (Bulmer, 1979, p. 6):
'We must remember that the probability of an event is not a quality of the event
itself, but a mere name for the degree of ground which we ... have for expecting
it ... Every event is in itself certain, not probable: if we knew all, we should either
know positively that it will happen, or positively that it will not. But its probability
means to us the degree of expectation of its occurrence, which we are warranted
in entertaining by our present evidence.' This would seem to be a useful definition
for actuaries and insurance brokers.

In science you should always follow the approach that best suits you. In the first,
a priori, view of probability, we simply **define** an unbiased coin as one for which the
probability of heads is 0.5 and use our laws of probability to work out whether
500 737 heads in one million throws is a reasonable outcome given our hypothesis
that the coin is unbiased. In other words, we make an hypothesis, do an experiment and
then see whether the resulting data are in reasonable agreement with our expectations:
if they are not we must reconsider the hypothesis and possibly our data as well. As
we shall see in this book, we do not sit around looking blankly at our data and then
suddenly jump up, saying, 'Eureka! tsetse flies like blue traps.' Usually we have a
pretty good idea of what we think might happen and we are more likely to say, 'I
wonder if the colour of the traps affects the number of tsetse flies that we catch?
Let us catch as many flies as we can with blue traps and yellow traps, and then see
whether the data we collect supports our hypothesis that the catch is affected by
the colour of the trap.' In this book we will talk about **testing hypotheses**. This not
only makes for more rigorous statistics but helps us to remember that the reason we
apply mathematics to biology is not to generate large amounts of obscure numbers,
but to help to formulate clear and precise hypotheses to test.

Sometimes the second, the long-run frequency idea, is the most useful: if you are
struggling to decide on how to assign frequencies or probabilities, or indeed to decide

on what might be the important factors in an experiment, you can always ask: What would I expect to happen if I repeated this experiment many times? We usually hope that the results of our experiments on new insect trap designs, improved insecticides, or high yielding strains of wheat, for example, will be used by other people in other places and this might help you to think about all of the possible things that might happen when the experiment is repeated many times.

The third, casino, approach can also be useful because it reminds us that the random nature of our observations arises, at least in part, from our lack of knowledge of the precise factors affecting our experiments. Given some information, we might then make certain predictions concerning the outcome of an experiment. But more information might lead us to change our prediction and this reflects well the provisional nature of scientific knowledge.

2.2 THE LAWS OF PROBABILITY

Once we have agreed on some concept of probability, we need to develop our notation. We could make statements, such as, 'If I toss an unbiased coin, the probability that it will land heads is 50%.' But this is long, and we will want to discuss considerably more complicated situations. So let us use the letter P for probability and define our **scale** of probability in such a way that **anything that is certain to happen has a probability of 1 and anything that is certain not to happen has a probability of 0**. We will then say, for example, that if we toss an unbiased coin,

$$P(H) = 0.5, \qquad\qquad 2.1$$

which is read, 'The probability of getting heads is 0.5'. Sometimes we will want to include explicitly the conditions under which our statement holds, and then we use a '|' (bar) to indicate a conditional probability and write instead

$$P(H|\text{unbiased coin}) = 0.5, \qquad\qquad 2.2$$

which reads 'The probability of getting heads, given that the coin is unbiased, is 0.5'. In another situation we might then write

$$P(H|\text{double-headed}) = 1, \qquad\qquad 2.3$$

which reads, 'The probability of getting heads, given that coin is double-headed, is 1'.

We can combine numbers in two important ways: we can add them and we can multiply them. The next thing we must do is to determine the equivalent laws for combining probabilities.

2.2.1 The law of addition

I think you will agree that if we spin an unbiased coin

$$P(H) = P(T) = 1/2. \qquad\qquad 2.4$$

And I think you will also agree that with an unbiased die the probability of throwing any number between 1 and 6 is 1/6, that is

$$P(1) = P(2) \ldots = P(6) = 1/6. \qquad 2.5$$

So if I were to throw a die and ask you to guess the number that turns up, you would consider that the probability of guessing correctly is 1 in 6, of guessing incorrectly is 5 in 6. You should demand odds of at least 5 to 1 to make it worth your while playing the game.

So far so good. But suppose I throw a die and let you guess any two numbers and we agree that if **either** of them comes up you win. What odds would then make it worth your while to play? For example, you might now bet that the result will be either a 2 or a 5, and so we want to know the probability of getting a 2 or a 5, or in our new notation, we want to know the value of $P(2$ or $5)$.

Let us examine the problem more carefully. When we throw an unbiased die, the reason for concluding that $P(4) = 1/6$ is as follows: There are six possible outcomes, $1, 2, \ldots 6$. Since we assume that each outcome is equally likely, we argue that the probability of getting 4, say, is simply equal to the number of ways we can get 4 (1) divided by the total number of ways the die can fall (6).

Suppose then that we throw a die and are allowed two bets. If you bet on 2 or 5, say, there are two ways in which you can win—namely, if the die falls 2 or 5—and four ways in which you can lose—namely, if the die falls 1, 3, 4 or 6. The total number of outcomes is still 6, so that the chance of winning is 2 out of 6 (and of losing is 4 out of 6) and we can write

$$P(2 \text{ or } 5) = 2/6 = 1/3. \qquad 2.6$$

Clearly, if you are going to play this game you should demand odds of at least 2 to 1. Now if you think about this argument, it should be clear that we have simply added up the number of (equally likely) ways in which you can win and divided that number by the total number of (equally likely) ways the game can come out, so that we can generalize Equation 2.6 and write

$$P(2 \text{ or } 5) = (1 + 1)/6 = 1/6 + 1/6 = P(2) + P(5) = 1/3, \qquad 2.7$$

since adding the number of ways we can win and then dividing that number by the total number of ways the dice can fall is just the same as adding the probabilities of each outcome. But we must be careful: if you bet two numbers but both bets are for 2, that is, if you make the silly bet that the outcome will be 2 or 2, then the probability of a win is $P(2$ or $2) = 1/6$, **not** 1/3, so that the rule only applies if the alternatives are **mutually exclusive**: that is to say, either of the two outcomes might happen but certainly not both.

Let us then summarize our **law of addition. For two mutually exclusive events, A and B, the probability that either A or B will occur is equal to the sum of their separate probabilities**: that is,

$$P(A \text{ or } B) = P(A) + P(B). \qquad 2.8$$

Indeed, we have come full circle, for we effectively used our law of addition to derive the results given in Equations 2.4 and 2.5. If the probability of getting heads is equal to the probability of getting tails, and if our law of addition holds so that they add up to 1, then each of them must be equal to 0.5.

We can now use this law to take another look at Mendel's experiments on peas (Mendel, 1866). Mendel crossed homozygous TT peas with homozygous tt peas, where T indicates the gene for tallness and t the gene for shortness. Each plant in the first filial generation, F_1, receives one gene from each of its parents and so they must all have one T and one t gene. Thus the F_1 plants are all heterozygous, with genetic constitution Tt. Since all of the plants in the first filial generation were tall, the gene for tall plants is dominant over the gene for short plants. Mendel then cross-fertilized these heterozygous plants and found that about 3/4 of them were tall and 1/4 were short. In the second filial generation, F_2, there are four ways in which the F_2 plants can inherit genes from the male and female F_1 parents, as indicated in Table 2.1.

Since each of the four outcomes in Table 2.1 is equally likely and the tall gene is dominant, the ratio of tall to short plants should be 3 to 1. We have effectively used the law of addition to determine the predicted ratio of tall to short plants since

$$P(F_2 \text{ is tall}) = P(TT) + P(Tt) + P(tT) = 1/4 + 1/4 + 1/4 = 3/4, \qquad 2.9$$

while

$$P(F_2 \text{ is short}) = P(tt) = 1/4. \qquad 2.10$$

There are three important points to note. First of all, the four outcomes in Table 2.1 are equally likely only if the F_2 plants are just as likely to inherit T genes as they are to inherit t genes. When we use statistics, we must be aware of the assumptions that underlie our analyses, especially when the assumptions are not explicitly stated, so that if the data do not agree with our expectation we have some idea of where to look for an explanation. The second point is that Tt and tT are distinguishable genetic constitutions: in the first case T came from the male and t from the female, while in the second case the reverse was true. We must be careful to distinguish events which, though seemingly the same, are in fact different. Lastly, we can apply the law of addition because the outcomes are mutually exclusive. For example, we could have TT **or** Tt, but we could not possibly have TT **and** Tt in the same plant.

In the previous chapter we saw that Mendel's tall and short plants did occur in the ratio of about 3 to 1 in the F_2 generation. However, being a good scientist, he

Table 2.1 The four possible (equally likely, mutually exclusive) genetic constitutions of F_2 peas produced from heterozygous F_1 males and females

F_1 male	F_1 female	F_2
T	T	TT
T	t	Tt
t	T	tT
t	t	tt

Table 2.2 Mendel's data for pairs of characters as they occurred in the F_2 generation of peas. The table gives the number of plants showing the dominant character, the number of plants showing the recessive character and the ratio of the two. The last row gives the total number of plants, showing the dominant or recessive character and the ratio of these two numbers

Character	Dominants	Recessives	Ratio
Round vs wrinkled seeds	5 474	1850	2.96
Yellow vs green seeds	6 022	2001	3.01
Purple vs white flowers	705	224	3.15
Smooth vs constricted pods	882	299	2.95
Axial vs terminal flowers	651	207	3.14
Green vs yellow unripe pods	428	152	2.82
Tall vs dwarf stems	787	277	2.84
Total	14 949	5010	2.98

did not stop there but went on to study several other pairs of characteristics, some of which are given in Table 2.2 (Mendel, 1866). The ratios of the numbers of peas having the dominant character to the numbers having the corresponding recessive character cluster around a value of 3, in support of Mendel's theory, although none of the numbers is precisely 3. The data for yellow and green seeds give a ratio very close to 3 while the data for green and yellow unripe pods give a ratio less close to 3. In later chapters we shall examine in detail the question 'How close is close?', and we shall develop statistical tests that will enable us to make precise statements about the meaning of 'closeness'. The theory that allows us to say just how close is close was not widely understood in Mendel's day, but he stated this aspect of the problem clearly: 'The true ratios of the numbers can only be ascertained by an average deduced from the sum of as many single values as possible; the greater the number, the more are merely chance effects eliminated' (Mendel, 1866). This is in essence a statement of our long-run frequency definition of probability. We see from Mendel's data that the ratio of the total number of dominants to the total number of recessives is very close to 3:1.

2.2.2 Conditional probability

Before we can establish the second important law of probability, we need to introduce the concept of **conditional probabilities**. We have agreed that if I throw one die, the probability that it will come up showing 2, say, is 1/6, which we can write as

$$P(2) = 1/6. \qquad 2.11$$

Suppose I now throw a die and look at it but hide it from you. I tell you that it has come up with an even number. What would you say is the probability that it is showing 2? Since there are now only three possible outcomes—2, 4 and 6—the probability that it is showing 2 is 1/3, which we can write as

$$P(2|\text{even}) = 1/3, \qquad 2.12$$

and Equation 2.12 reads 'The probability of getting a 2, given that the number is even, is 1/3'.

We can now use the notion of conditional probability to define another important concept, that of **statistical independence**. Suppose we throw two dice, which we call a and b. Then the probability that a shows 2 has nothing at all to do with the number that b shows and it **must** be the case that

$$P(a \text{ shows } 2 | b \text{ shows even}) = P(a \text{ shows } 2) = 1/6, \qquad 2.13$$

while we know that the probability that a shows 2 **does** depend on whether or not a shows even, since

$$P(a \text{ shows } 2 | a \text{ shows even}) = 1/3 \neq P(a \text{ shows } 2) = 1/6. \qquad 2.14$$

These observations provide us with an important test of statistical independence, because if we find that

$$P(A|B) = P(A), \qquad 2.15$$

we can conclude that A and B are statistically independent. We should also note that in general

$$P(A|B) \neq P(B|A). \qquad 2.16$$

This is easily seen with our die since $P(2|\text{even}) = 1/3$, while $P(\text{even}|2) = 1$.

2.2.3 The law of multiplication

Our law of addition gave us a rule for calculating $P(A \text{ or } B)$. We now want a rule for calculating $P(A \text{ and } B)$. For example, we might want to throw two dice and work out the probability that they will both show 2. We have already seen (Equations 2.9 and 2.10) that Mendel's theory predicts that for peas in the second filial generation $P(\text{plants are tall}) = 0.75$, $P(\text{plants are short}) = 0.25$, $P(\text{seeds are round}) = 0.75$, and $P(\text{seeds are wrinkled}) = 0.25$. How then will we calculate $P(\text{plants are tall and seeds are round})$ or $P(\text{plants are short and seeds are wrinkled})$?

For any two outcomes, A and B, we want an expression for $P(A \text{ and } B)$. We can consider this problem in two stages. First, what is the probability that we will get B? Second, given that we do get B, what is the probability that we will then get A also? The answer to the first question is simply $P(B)$ and the answer to the second question is the conditional probability $P(A|B)$. I now want to convince you that to determine $P(A \text{ and } B)$ we must **multiply** these two probabilities together, so the **law of multiplication** states: **the probability of getting A and B is the probability of getting A, given B, times the probability of getting B.**

$$P(A \text{ and } B) = P(A|B) \times P(B). \qquad 2.17$$

In words, the probability of getting A and B is the probability of getting B times the probability of getting A given that we already have B. If A and B are statistically independent so that $P(A|B) = P(A)$, as in Equation 2.15, then Equation 2.17 becomes

$$P(A \text{ and } B) = P(A) \times P(B). \tag{2.18}$$

Let us examine this simpler version first.

Suppose we throw two dice, one coloured red and one coloured blue so that we can tell them apart. I think we agree that, unless we glue them together, what happens to the red one will not affect what happens to the blue one, so that they are statistically independent. This time we will play a game in which you have to guess a number for each die. Suppose you guess 2 for red and 5 for blue; what then is the probability that the red will show 2 and the blue will show 5 and you will win? That is, we want to evaluate $P(2 \text{ on red and } 5 \text{ on blue})$.

Since the number of possible outcomes is small, we can write them all down as in Table 2.3 and we see that there are $6 \times 6 = 36$ ways in which the dice can fall. Out of the 36 possible outcomes, the red shows 2 on six occasions, so that the probability that the red shows 2 is $6/36 = 1/6$, just as we expect. We now look at these six outcomes, the second row of Table 2.3, and on only one of them does the blue die show 5. So the probability that the blue die shows 5, given that the red die shows 2, is $1/6$. Again, this is just as we expect, since the number that shows on the blue die does not depend on the number that shows on the red die. We then see that on $1/6$ of $1/6$ of the throws, the red shows 2 and the blue shows 5, so that the probability that the red shows 2 and the blue shows 5 is $(1/6) \times (1/6) = 1/36$, in agreement with Equation 2.18, which is the law of multiplication for statistically independent events.

To illustrate the law of multiplication for events that are not statistically independent, imagine that we throw a die and ask: what is $P(\text{even and} \leqslant 5)$? There are two outcomes, namely 2 and 4, that are both even and less than or equal to 5 so the answer must be $1/3$. To confirm that the law of multiplication gives the same answer we note that

$$P(\text{even and} \leqslant 5) = P(\text{even}| \leqslant 5) \times P(\leqslant 5) = (2/5)(5/6) = 1/3 \tag{2.19}$$

Table 2.3 The 36 ways in which two dice can fall. The numbers in the table give the sum of the numbers on the dice

		Blue					
		1	**2**	**3**	**4**	**5**	**6**
	1	2	3	4	5	6	7
	2	3	4	5	6	7	8
Red	**3**	4	5	6	7	8	9
	4	5	6	7	8	9	10
	5	6	7	8	9	10	11
	6	7	8	9	10	11	12

or, if you prefer,

$$P(\text{even and} \leqslant 5) = P(\leqslant 5 \mid \text{even}) \times P(\text{even}) = (2/3)(1/2) = 1/3. \qquad 2.20$$

In a case as simple as this we would not bother to use the law of multiplication and would simply write down the answer directly, but in more complicated situations we will have to use the law of multiplication. When faced with a statistical result whose meaning may not be obvious, it is always good practice to think of a simple example, such as the one above, and put in some numbers to help you to understand the result.

We can now look again at the casino game we suggested in section 1.1.1, in which we throw two dice and have to guess what the sum of the two numbers will be. From Table 2.3 it is clear that we must guess a number between 2 and 12 and that the dice can fall in 36 different ways. We can also see by counting the number of ways various sums can turn up that $P(2) = 1/36$, $P(3) = 2/36, \ldots, P(7) = 6/36$, $P(8) = 5/36$, and so on. To see how we can apply our laws to reach the same conclusion, we note that we will get 5, for example, if we throw 1 and 4, 2 and 3, 3 and 2, or 4 and 1. We then have

$$P(5) = P[(1 \text{ and } 4) \text{ or } (2 \text{ and } 3) \text{ or } (3 \text{ and } 2) \text{ or } (4 \text{ and } 1)]. \qquad 2.21$$

Using the law of addition we can write out the 'or's so that

$$P(5) = P(1 \text{ and } 4) + P(2 \text{ and } 3) + P(3 \text{ and } 2) + P(4 \text{ and } 1). \qquad 2.22$$

We can then use the law of multiplication (for independent events) to write out the 'and's, so that

$$P(5) = P(1) \times P(4) + P(2) \times P(3) + P(3) \times P(2) + P(4) \times P(1). \qquad 2.23$$

Since each of the individual outcomes has a probability of 1/6,

$$P(5) = (1/6) \times (1/6) + (1/6) \times (1/6) + (1/6) \times (1/6) + (1/6) \times (1/6) = 4/36. \quad 2.24$$

In a case as simple as this, it is easy to count the number of successes and the total number of possible outcomes and divide the one by the other, and that is all that our laws are doing for us. But if we have hundreds of possible outcomes, it quickly becomes tedious if not impossible to enumerate all the possibilities; we then use our laws to help us manipulate probabilities.

Now we can work out the best strategy for our casino game. Remember that you pay £1 every time you guess wrong and you receive as many pounds as the number you guess if you guess right. If you bet 5, say, your probability of wining is 4/36 and since you are then paid £5 your average winnings in 36 games are £20. Since you pay £1 each time that you make a wrong guess and your probability of losing is 32/36, your average losses in 36 games are £32. Table 2.4 summarizes the results for some of the bets that you might make. Clearly, your best bet is 7 although 8 is almost as good and you should also do well if you bet 9.

Let us return again to Mendel and his peas. We have seen that Mendel was able to identify a number of dominant—recessive pairs of characters. This led him to

Table 2.4 The gains and losses in pounds for various bets in the casino game described in section 1.2.1

Bet	P(winning)	Gain/36 games	Loss/36 games	Net gain
5	4/36	20	32	− 12
6	5/36	30	31	− 1
7	6/36	42	30	+ 12
8	5/36	40	31	+ 9
9	4/36	36	32	+ 4
10	3/36	30	33	− 3

wonder if the pairs of characters were linked in any way. For example, genes for round seeds are dominant over genes for wrinkled seeds, while genes for yellow seeds are dominant over genes for green seeds; perhaps the round seeds will tend also to be yellow while the wrinkled seeds will tend to be green. We can rephrase this using our newly developed terminology and ask: is $P(\text{yellow}|\text{round}) = P(\text{yellow})$? is $P(\text{wrinkled}|\text{yellow}) = P(\text{wrinkled})$? and so on. In other words, we can test the pairs of characteristics to see if they are statistically independent. To do this, Mendel proceeded as in his earlier experiments, but this time he crossed a pure line having round yellow seeds with a pure line having wrinkled green seeds. All the resulting hybrids had round yellow seeds, confirming the dominance of the round and yellow characters. When these hybrids were in turn crossed, he obtained the results shown in Table 2.5 from which we can calculate the proportions in which the various characteristics occur. For example, we have

$$\hat{P}(\text{round}) = 423/556 = 0.761 \quad \hat{P}(\text{wrinkled}) = 133/556 = 0.239$$

$$\hat{P}(\text{round}|\text{yellow}) = 315/416 = 0.757$$

$$\hat{P}(\text{wrinkled}|\text{yellow}) = 101/416 = 0.243. \qquad 2.25$$

(I have used a circumflex '^' to indicate a probability estimated as the ratio of two frequencies to distinguish it from the 'true' underlying probability.) First of all we see that about 3/4 of all the seeds are round. More importantly, if we consider only the yellow seeds, about 3/4 of them are round and if we consider only the green seeds, about 3/4 of them are round. Writing this out formally we have

$$\hat{P}(\text{round}) = 0.761 \approx \hat{P}(\text{round}|\text{yellow}) = 0.757 \approx \hat{P}(\text{round}|\text{green}) = 0.771. \qquad 2.26$$

Table 2.5 Mendel's data on the distribution of seed colour and shape

	Yellow	Green	Total
Round	315	108	**423**
Wrinkled	101	32	**133**
Total	**416**	**140**	**556**

Table 2.6 Bateson's data on the distribution of pollen shape and the colour of the flowers in the F_2 generation

	Purple flowers	Red flowers	Total
Long pollen	1528	117	**1645**
Round pollen	106	381	**487**
Total	**1634**	**498**	**2132**

So the estimated probability that a seed will be round is about the same whether the seed is green or yellow and the traits for colour and shape are statistically independent (Equation 2.15).

With the rediscovery of Mendel's work, William Bateson repeated many of Mendel's experiments (Bateson, 1909). In one experiment Bateson started from two pure strains of peas, the first having long pollen and purple flowers, the second having round pollen and red flowers. In the F_1 generation all the plants had long pollen and purple flowers, these being the dominant characters. In the F_2 generation he obtained the results shown in Table 2.6.

If we go through the same exercise as we did for Mendel's data, we find that

$$\hat{P}(\text{long pollen}) = 1645/2132 = 0.772$$

$$\hat{P}(\text{round pollen}) = 487/2132 = 0.229 \qquad 2.27$$

in about the proportions expected on the basis of Mendel's theory. However,

$$\hat{P}(\text{long pollen}|\text{purple flowers}) = 1528/1634 = 0.935$$

$$\hat{P}(\text{long pollen}|\text{red flowers}) = 117/498 = 0.215. \qquad 2.28$$

In this case, the two pairs of characters are not statistically independent, since

$$\hat{P}(\text{long pollen}) = 0.772$$

$$\neq \hat{P}(\text{long pollen}|\text{purple flowers}) = 0.935 \qquad 2.29$$

$$\neq \hat{P}(\text{long pollen}|\text{red flowers}) = 0.215,$$

so that the probability that a pea has long pollen does depend on the colour of its flowers.

Two points are worth noting. First of all, we have said that 0.935 and 0.215 in Bateson's experiments differ from 0.772 but that 0.757 and 0.771 in Mendel's experiments are not different from 0.761. Clearly Mendel's numbers are in closer agreement with each other than are Bateson's but we are going to have to decide when differences are and are not significant. The second point is that the statistics merely suggest that something interesting might be going on since flower colour and pollen shape seem to be related while seed colour and seed shape do not. But the statistics tell us nothing at all about the biology. We now know that the reason for these results is that genes are carried on chromosomes and if two pairs of genes are found on the same chromosome they are likely to be inherited together, and therefore linked, while if they are found on different chromosomes they are likely to be inherited independently of one another. In other words, the statistics are important and suggestive and give us hints as to where to look, but it is up to us to interpret our data and draw out the biological implications.

2.3 SUMMARY

Probability is a subtle concept and it can be thought about in several different ways: as something that is given, as the outcome of many repeats of the same experiment or as a measure of our confidence or degree of belief in a certain event. How you think about it will depend on you and on the problem at hand.

In order to progress we need to be able to calculate the probability that several different things will happen in an experiment and the laws for combining probabilities are central to everything that follows. I have not tried to present them in a rigorous way; that has been done by others (Freund, 1972). What matters is that you know how and when to apply the two laws for combining probabilities and the conditions under which they hold and that you understand clearly when events are mutually exclusive and the meaning of statistical independence.

We have already seen that even in simple experiments, such as the throwing of two dice, the probabilities of each of the possible (combined) outcomes occurring may differ. In the next chapter we will consider the probability distributions of the outcomes of various experiments and think about ways to represent these distributions.

2.4 EXERCISES

1. Suppose that you throw one die. Write down: $P(1)$, $P(1$ or $2)$, $P(1$ or 2 or $3)$, $P(1$ or 2 or 3 or $4)$, $P(1$ or 2 or 3 or 4 or $5)$, $P(1$ or 2 or 3 or 4 or 5 or $6)$. Write down $P(\text{not } 5)$, i.e. the probability that it does not show 5.

2. Suppose that you throw two dice, a and b. Write down: $P(a$ shows 1 and b shows 2$)$, $P(a$ shows 1 and b shows 2 or 3$)$.

3. Use Mendel's data in Table 2.5 to estimate $P(\text{round}|\text{green})$, $P(\text{wrinkled}|\text{green})$, $P(\text{yellow})$, $P(\text{yellow}|\text{round})$, $P(\text{yellow}|\text{wrinkled})$, $P(\text{yellow and round})$.

4. Use Bateson's data in Table 2.6 to estimate $P(\text{round pollen})$, $P(\text{round pollen}|\text{purple flowers})$, $P(\text{round pollen}|\text{red flowers})$, $P(\text{purple flowers})$, $P(\text{purple flowers}|\text{round pollen})$, $P(\text{purple flowers}|\text{long pollen})$.

5. Use Bateson's data in Table 2.6 to estimate $P(\text{long pollen and purple flowers})$, $P(\text{round pollen and red flowers})$.

6. Suppose that you throw three coins. How many ways can they fall so as to give (a) three heads, (b) two heads, one tails, (c) one heads, two tails and (d) three tails? If the coin is unbiased, what is the probability of each outcome (a) to (d)?

7. Suppose that we have four plants in the F_1 generation as bred in Mendel's experiment so that one has genetic constitution TT, two Tt, and one tt. Each of these four plants is then crossed with each of the others. Suppose that each cross produces four new plants. Calculate the expected number of plants with each genetic constitution in the next generation. This illustrates the Hardy-Weinberg law (Roughgarden, 1979), that in the absence of natural selection and with random mating the genotypic proportions remain fixed. [Hint: You can solve this problem either by enumerating all possible crosses or by using the laws of probability.]

3

Representations

If we fully apprehend the pattern of things of the world will it not be found that
everything must have a reason why it is as it is? And also a rule [of co-existence with
all other things] to which it cannot but conform? Is not this just what is meant by Pattern?
Hsü Hêng (1209–1281) (Needham, 1972, p. 38)

When we throw a die we cannot predict with certainty which number will turn up.
In the nineteenth century scientists would have argued that our inability to predict
the outcome reflected our lack of knowledge, so that if we knew precisely how we
had thrown the die, the precise details of its shape and weight, how each tiny current
of air was moving, and so on, then we could, in principle, tell with certainty how it
would land. This was the essence of Mill's second statement of probability (section
2.1). Nowadays, uncertainty is regarded as being more fundamental than was pre-
viously thought, and quantum mechanics is based on laws that are themselves
probabilistic. Einstein never accepted quantum mechanics as a complete theory of
matter because of the probabilistic nature of its laws (Jammer, 1974). 'Quantum
mechanics is very imposing,' he wrote, 'but an inner voice tells me that is still not
the true Jacob. The theory yields much, but it hardly brings us nearer to the secret
of the Old One. In any case I am convinced that He does not throw dice.' Realizing
that combining quantum mechanics and general relativity in the study of black holes
leads to even greater uncertainty, Stephen Hawking comments (Boslough, 1984):
'God not only plays dice, but he sometimes throws them where they cannot be seen!'
Stewart (1990) in his book on Chaos now adds: 'For we are beginning to discover
that systems obeying immutable and precise laws do not always act in predictable
and regular ways. Simple laws may not produce simple behaviour. Deterministic laws
can produce behaviour that appears random. Order can breed its own kind of chaos.
The question is not so much *whether* God plays dice, but *how* God plays dice.'

The discovery of chaotic behaviour, and its application to biology by Robert May
(1976) and others, shows that even entirely deterministic classical systems can be
inherently unpredictable and John Conway's delightful game 'Life' illustrates deterministic but unpredictable dynamics (Gardiner, 1970, 1971, 1983). It is believed that
weather is chaotic so that no matter how powerful our computers become, we will
never be able to make precise weather forecasts more than a week or so into the
future. In later chapters we will have to consider the degree to which the 'randomness'

in our data reflects essentially unknowable features of our experiments and the extent to which it reflects aspects of the experiment we have not understood or taken into account.

If we throw a die many times and get 6, 3, 5, 3, 2, 5, 1,..., say, these give us a sequence of random numbers. Now, although we cannot say with certainty which number will turn up next, we can make precise statements about the **probability** that each number will turn up next. In other words, the **probability distribution** of the outcome of even a single throw of a die can be known precisely. In our study of statistics we are therefore concerned with determining the probability distributions from which the numbers that we measure in such experiments come.

3.1 DISCRETE RANDOM VARIABLES

In statistics we are concerned not with the particular results of individual measurements but with the distribution of the measured values. A great deal of work in statistics is spent in identifying and describing the distribution associated with a particular set of measurements or observations and the first thing we must do is to consider ways of representing distributions of random numbers.

3.1.1 Frequency distributions

In the previous chapter we saw that when we throw a die, the probability of getting any number between 1 and 6 is 1/6. To illustrate this, I used a computer to simulate an experiment in which I threw a die 600 times, getting 95 ones, 101 twos, 107 threes, 112 fours, 91 fives and 94 sixes. As expected, each number came up about 100 times or about one time in six. However, it is easier to see what is going on if we plot the **frequency distribution** of outcomes in the form of a **bar chart** in which we draw a series of boxes in such a way that the height of each box gives the number of times

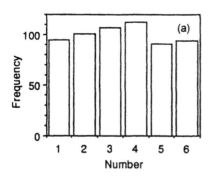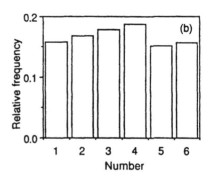

Figure 3.1 The result of a computer simulation of an experiment in which a die is thrown 600 times. (a) The frequency distribution (b) the relative frequency distribution of the possible outcomes. The two plots differ only in the choice of scale on the vertical axis.

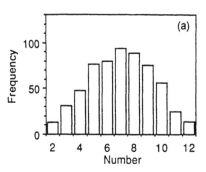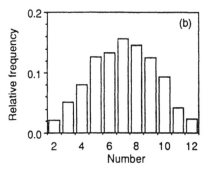

Figure 3.2 The result of a computer simulation of an experiment in which two dice are thrown 600 times and the pairs of numbers are added together. (a) The distribution of outcomes and (b) the relative frequency distribution. The two plots differ only in the choice of scale on the vertical axis.

the result for that box turns up. This is shown in Fig. 3.1a from which we see that each number occurs with about the same frequency.

An alternative, but equivalent, way to plot the data involves dividing the number of times each outcome occurs by the total number of events. This is called the **relative frequency** distribution and is shown in Fig. 3.1b. The frequency distribution shows us that we get a 1 in 95 out of 600 throws; the relative frequency distribution shows us that we get a 1 with probability 0.16.

To illustrate this further I threw two (simulated) dice 600 times, added together each pair of numbers that turned up and then plotted the frequency distribution of the sums. In Fig. 3.2 we can see that the two numbers add up to 7, for example, in 95 out of 600 or 16% of the throws.

3.1.2 Probability distributions

Now we know how to calculate the probabilities of each outcome in each of the two experiments shown in Figs 3.1 and 3.2. We believe that the more times we throw the dice the closer the observed relative frequencies will be to the theoretical probabilities, which are plotted in Fig. 3.3. We can regard each of the relative frequency distributions plotted in Figs 3.1 and 3.2 as estimates of the underlying 'true' probability distributions shown in Fig. 3.3.

In passing we should note that these are examples of **mathematical models**. In Fig. 3.3 we have **analytical models** in which we calculate the probabilities of various outcomes from first principles; because the problems are so simple, we can do this exactly. In biology we are seldom able to carry out exact calculations and so we often make **simulation models**, which mimic our experiments, or at least what we believe to be the most important aspects of our experiments. This is what we did to obtain the results shown in Figs 3.1 and 3.2. We could then do an actual experiment by throwing two real dice 600 times and comparing the outcome with the results of

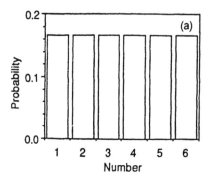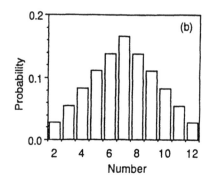

Figure 3.3 The theoretical probability distribution functions corresponding to the two experiments shown in Figures 3.1 and 3.2. (a) The probability distribution for the number that shows when one die is thrown and (b) the probability distribution for the sum of the numbers that show when two dice are thrown.

our analytical (Fig. 3.3) or simulation (Fig. 3.2) models in order to decide if there is evidence for bias in our dice.

3.1.3 Cumulative distributions

Bar charts give us a graphical representation of the frequency or the probability with which a particular outcome turns up in an experiment. However, instead of asking 'What is the probability of getting a 3?', we might ask, 'What is the probability of getting a number **less than or equal to** 3?' The latter question is important because when we test an hypothesis against our data, we are usually looking for extreme cases or unusual results and we want to know the probability of getting a result as big as or bigger than (or as small as or smaller than) the one we actually observe. For example, when we vaccinated the children against polio (section 1.1.3) we wanted to know if the number who then developed polio was unusually small. If we are testing an insect trap, we want to know if it catches an exceptionally large number of insects.

If we throw one die, then using the law of addition we have

$$P(1) = 1/6 \qquad\qquad 3.1$$

$$P(1 \text{ or } 2) = P(1) + P(2) = 1/3 \qquad\qquad 3.2$$

$$P(1 \text{ or } 2 \text{ or } 3) = P(1) + P(2) + P(3) = 1/2 \qquad\qquad 3.3$$

and so on. In other words, to obtain the probability that the result is less than or equal to 3, say, we add the probabilities of each outcome from 1 to 3. Now we could always write these sums out in full. But I am sure you will agree that if we had 100 possible outcomes, writing out the sums would rapidly become tedious. So we introduce a new notation in which we use the Greek capital letter Σ, 'sigma', the equivalent of our S, to represent summing over various outcomes. We can then

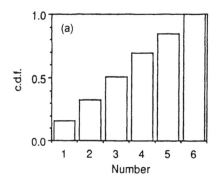 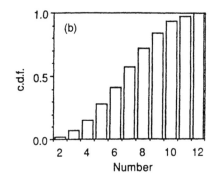

Figure 3.4 The cumulative distribution functions, c.d.f.s, corresponding to the relative frequencies given in Figures 3.1b and 3.2b.

rewrite Equation 3.3 more compactly as

$$P(1 \text{ or } 2 \text{ or } 3) = \sum_{i=1}^{3} P(i). \qquad 3.4$$

The right-hand side of Equation 3.4 reads: 'the sum of $P(i)$ from $i = 1$ to 3'. Since Equation 3.4 gives us the accumulated probability for all outcomes less than 3, we call it the **cumulative distribution function** or c.d.f. Writing it with a C, we have

$$C(3) = \sum_{i=1}^{3} P(i) = 1/2 \qquad 3.5$$

for the cumulative probability of throwing one die and getting a number less than or equal to 3. Figure 3.4 gives the cumulative distributions for the data given in Figs 3.1 and 3.2. From the cumulative distributions we can read off directly the probability that a number is less than or equal to any chosen value. The c.d.f. must always lie between 0 and 1.

3.2 CONTINUOUS RANDOM VARIABLES

So far we have talked about experiments in which we measure **discrete variables**, which can take on discrete values only, such as heads or tails, purple or white, and 1 or 2 or 3. We will discuss other discrete variables such as the number of armyworms in a field of maize or the number of ticks feeding on different rabbits. But we will also measure **continuous variables**, which can take any value on a continuum of numbers, such as the wing length of a fly or the weight of an engorged tick and we will need to see how these fit into our scheme of things.

Fortunately, the ideas we have developed to describe discrete distributions all have equivalents for continuous distributions and if we are clear about the one we should be clear about the other. There is, however, an entire branch of mathematics that deals with changes in continuous functions. This was developed in the eighteenth century by Isaac Newton in England and Gottfried Leibniz in France. This study is called calculus: differential calculus enables one to find the slope of a continuous curve at any point on it and integral calculus enables one to find the area under a continuous curve between any two points. We will not be able to develop and use calculus in this book; we will be able to do all that we need without it. However, for those of you who are familiar with calculus, remember that many of the results we will obtain could be more easily obtained using the calculus.

3.2.1 Frequency distributions

As an example of a continuous distribution, consider the data shown in Table 3.1, which give measurements in millimetres of the lengths of the wings of 100 houseflies (Sokal and Hunter, 1955). In Table 3.1 we see that there are lots of 4.5s and only one 3.6, but a graphical representation should make it easier to see how the lengths of the wings are distributed. If the wing lengths occurred in discrete categories, as in our earlier examples, we could plot them as a bar chart, which would indicate how the lengths of the wings are distributed. We can still do this, even though the wing lengths are continuously distributed, but we will have to create our own discrete categories. For example, we can make a series of categories by taking 0.1 mm ranges. Since the shortest wing is 3.63 mm long, let us take wing lengths from 3.6 to just less than 3.7 mm for our first category, which in this example contains one fly, from 3.7 to just less than 3.8 mm for the second category, containing one fly, from 3.8 to just less than 3.9 mm for the third category, containing two flies, ..., from 4.5 to just less than 4.6 mm for the 26th category, containing 12 flies, and so on. We can then plot a bar chart, just as we did before and the result is shown in Fig. 3.5, which we call a **histogram**. Since each interval is 0.1 mm wide, I have divided each frequency by 0.1 so that the units are frequency per 1.0 mm rather than frequency per 0.1 mm. The difference between a bar chart and a histogram is that in the histogram we have chosen our own categories. (Many authors use 'bar chart' and 'histogram' interchangeably.) With a histogram we can make the categories as broad or as narrow as we

Table 3.1 Lengths in millimetres of the wings of 100 houseflies. The measurements have been arranged in order of increasing length

3.63	3.71	3.83	3.86	3.91	3.99	4.05	4.09	4.09	4.09
4.10	4.11	4.13	4.15	4.15	4.16	4.23	4.24	4.25	4.26
4.27	4.28	4.29	4.30	4.33	4.35	4.38	4.39	4.39	4.39
4.40	4.40	4.42	4.43	4.46	4.47	4.47	4.48	4.50	4.50
4.52	4.53	4.54	4.56	4.57	4.58	4.59	4.59	4.59	4.59
4.60	4.61	4.61	4.61	4.62	4.63	4.64	4.65	4.68	4.69
4.72	4.73	4.73	4.74	4.74	4.76	4.78	4.80	4.80	4.80
4.81	4.81	4.84	4.86	4.86	4.87	4.88	4.91	4.91	4.97
4.99	4.99	4.99	5.00	5.00	5.03	5.08	5.08	5.08	5.08
5.10	5.13	5.15	5.20	5.26	5.27	5.34	5.38	5.40	5.57

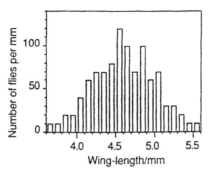

Figure 3.5 The wing lengths shown in Table 3.1 plotted as a histogram. The vertical axis gives the number of flies per millimetre whose wings fall into each 0.1 mm range.

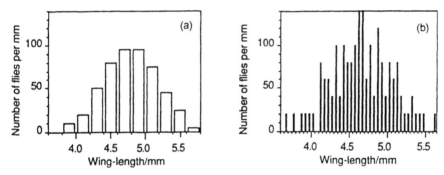

Figure 3.6 The histogram shown in Figure 3.5 replotted after (a) doubling and (b) halving the widths of the categories to which the wing lengths are allocated. By plotting the data as the frequency per millimetre, we keep the vertical axes the same in (a) and (b).

like, but unless we are very careful, we must keep the widths the same within any one histogram. One unfortunately common way of misrepresenting data is to vary the width of the categories to change the appearance of the histogram. An example of such misrepresentation is given in Exercise 3.4.3.

In Fig. 3.6 I have first doubled and then halved the widths of the categories in Fig. 3.5. Doubling the width of each category to 0.2 mm produces a smoother histogram but we are in danger of losing some of the details of the shape. Halving the width of each category to 0.05 mm reveals more structure but the statistical variation now makes it more difficult to pick out the underlying shape of the distribution.

By plotting the number of flies per unit of measurement on the vertical axis we are able to use the same vertical scale in all three cases in Figs. 3.5 and 3.6 and an important result follows: the area under any part of the histogram is equal to the number of flies in the corresponding range. For example, the height of the bar in Fig. 3.5 covering the range 5.0 to 5.1 mm is 70/mm while the width of the bar is 0.1 mm. The area of the rectangle, 70 × 0.1 = 7, gives the number of flies in the range covered by that bar. Finally, we can divide the area under any part of the histogram by the total number of flies to get the probability that the length of a fly's wing falls into a given range.

For discrete variables, we regarded each experimental frequency distribution as an approximation to an underlying 'true' probability distribution. What then is the equivalent of the 'true' theoretical probability distribution for continuous variables?

3.2.2 Probability density functions

The **probability density function**, or **p.d.f.**, is the continuous equivalent of the probability distribution for discrete variables. To illustrate the p.d.f., the data in Fig. 3.5 have been replotted in Fig. 3.7, after dividing by the total number of flies, so that the frequencies become probabilities. It then turns out that the **normal distribution** shown by the smooth curve in Fig. 3.7 is a good approximation to the p.d.f. It is called the normal distribution because it is the most common distribution; things are normally distributed in this way. Although the mathematical definition of the normal distribution function might seem obscure, it is the most important distribution in statistics, so we will digress for a while and consider it carefully.

We want a functional form for the bell-shaped curve in Fig. 3.7. Here we will not derive it from first principles, but if we plot a graph of $P(x)$ against x with

$$P(x) = e^{-x^2}, \qquad\qquad 3.6$$

where $e = 2.71828\ldots$ is the number that provides the base for natural logarithms, we will obtain a curve of the desired shape. When $x = 0$, $P(x) = 1$ and as x deviates from 0 in either direction, $P(x)$ becomes smaller. For very large values of x (in either direction) $P(x)$ becomes very small so that the curve flattens out along the x axis. Now we want to be able to scale the **height** of our curve to match our data so we

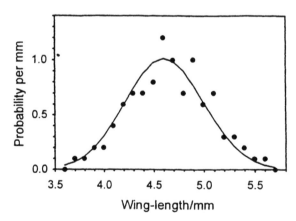

Figure 3.7 The data of Figure 3.5, after dividing by the total number of flies. The line is the probability density function for the normal distribution that best fits the data.

multiply the right-hand side of Equation 3.6 by a number N to get

$$P(x) = Ne^{-x^2}. \qquad 3.7$$

We also want to be able to **shift** our curve to match our data and we achieve this by subtracting a number m from x so that Equation 3.7 becomes

$$P(x) = Ne^{-(x-m)^2}. \qquad 3.8$$

In section 5.3 we will see that m is the mean of the distribution. We also want to be able to vary the **width** of our curve to match our data and we achieve this by dividing $x - m$ by a number s so that Equation 3.7 becomes

$$P(x) = Ne^{-(x-m)^2/2s^2} \qquad 3.9$$

and if we include the extra factor of 2 then, as we will see in section 5.3, s is the standard deviation of the distribution. Equation 3.9 is very important and you should plot graphs of $P(x)$ against x for various values of N, m and s to make sure that you understand the effect that each has on $P(x)$. The values that give the best fit of Equation 3.9 to the housefly data turn out to be 4.604 mm for m, 0.393 mm for s and 1.015/mm for N and these values were used to calculate the smooth curve in Fig. 3.7.

We have to think carefully about the meaning of a p.d.f., such as the normal distribution function. The continuous curve in Fig. 3.7 must be equivalent to the rectangles that we use for our histograms. Since the curve is continuous, we can imagine that it is made up of a very large number of very thin rectangles as in Fig. 3.8 where each rectangle is 0.1 mm wide. The area under the curve between any two values of x will then be equal to the area under the corresponding rectangles and therefore equal to the probability that the length of a fly's wing falls into a given range. Since the area under the entire curve, from $-\infty$ to $+\infty$ gives the probability that the length of a wing takes on any value, it must be 1 and to ensure that this is so the N in Equation 3.9 must be equal to

$$N = 1/s\sqrt{2\pi}. \qquad 3.10$$

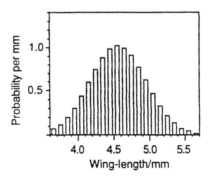

Figure 3.8 The normal distribution curve of Figure 3.7, which we can approximate by a series of narrow rectangles. In this plot the rectangles are 0.1 mm wide.

The concept of a probability density is central to much of what follows. First of all we note that whereas a probability is always dimensionless, the units of $P(x)$, in this example, are probability per mm, as expected for a 'density'. Since the height of each rectangle is equal to the probability per mm that a fly has a certain wing length, $P(x)$ gives the probability per unit length that a wing is x mm long. Suppose, for example, that we choose a wing length of 5 mm. $P(5)$ is equal to 0.67/mm for $m = 4.604$ mm and $s = 0.393$ mm (note the units). If we multiply $P(5)$ by 0.1 mm, the units cancel and the **probability** that the length of a wing lies within 0.1 mm of 5 mm, that is between 4.95 and 5.05 mm, is 0.067, or

$$P(4.95 \text{ mm} \leqslant \text{wing length} \leqslant 5.05 \text{ mm}) = 0.067. \qquad 3.11$$

If we count the number of flies in Table 3.1 whose wings are between 4.95 and 5.05 mm in length, we have seven flies, which means that the probability, estimated directly from the data, that the length of a wing is in this range is 0.07, in good agreement with the value obtained from the p.d.f. of 0.067. If we multiply $P(5)$ by 0.2 mm we have

$$P(4.9 \text{ mm} \leqslant \text{wing length} \leqslant 5.1 \text{ mm}) = 0.12. \qquad 3.12$$

We can again count the number of flies in Table 3.1 whose wings are more than 4.9 and less than 5.1 mm long, and we get 14 flies, so that the probability, estimated directly from the data, that the length of a wing lies in this range is 0.14, again in good agreement with the value obtained from the p.d.f. of 0.12. This is valid if the range we consider is sufficiently small, because we can then treat $P(x)$ as constant within this range and we are effectively working out the area under the curve between 4.95 and 5.05 mm in the first case and between 4.9 and 5.1 mm in the second[1].

The two important things to remember are:

- The probability density function or p.d.f. is the probability **per unit of measurement** that the measured variable has a particular value.
- The probability that a particular observation lies in a given range is equal to the area under the curve of the p.d.f. over this range and **if the range is small** we can calculate the area approximately by taking the value of the p.d.f. at the midpoint of the range and then multiplying by the range.

3.2.3 Cumulative distribution functions

Just as for discrete distributions, we want to be able to evaluate the probability that the length of the wing of a fly is less than 3.00 mm or greater than 4.00 mm, and so on. If we approximate the continuous probability density function by a histogram, we can proceed as we did in the case of discrete distributions and simply add up the probabilities per unit of length for all of the bars in the histogram below the critical level and then multiply this number by the width of the bars. If we do this for the data on houseflies, we obtain the **cumulative distribution function**, or **c.d.f.**

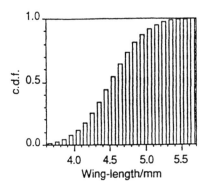

Figure 3.9 The cumulative distribution function (c.d.f.) for the housefly data shown in Figure 3.5.

shown in Fig. 3.9, corresponding to the histogram in Fig. 3.5. Formally we write

$$C(x_n) = \sum_{i=1}^{n} P(x_n)\delta x \qquad 3.13$$

where x_n labels the bars and δx is the width of the bars.

For continuous distributions the c.d.f. has several advantages over the p.d.f. First of all, whereas the p.d.f. is a **density** and gives the probability per unit of measurement that a certain result is observed, the c.d.f. is a **probability** and gives the probability that, in our example, the length of a housefly's wing is less than or equal to a set value. Secondly, although we can calculate the c.d.f. by first dividing the range of the observed variables into discrete categories, constructing a histogram and adding up the areas of successive histograms, there is a more direct way of obtaining the c.d.f. We simply take the length of the shortest wing, 3.63 mm, and we have one fly whose wing length is less than or equal to this. We then take the next smallest wing length, 3.71 mm, and we have two flies whose wings are less than or equal to this length. We proceed in this fashion up to the longest wing and then divide the number of each fly by the total number of flies to get the c.d.f. You see that we do not need to worry about setting up categories. This is illustrated in Table 3.2 and Fig. 3.10.

Table 3.2 The flies from which the data in Table 3.1 were obtained were ordered according to the lengths of their wings so that fly 1 has the shortest wing, fly 2 has the next shortest and so on. The c.d.f. is obtained by plotting the number of each fly in the ordered sequence divided by the total number of flies against the length of the wing for each fly as shown in Figure 3.10

Fly number	1	2	3	···	99	100
c.d.f.	0.01	0.02	0.03	···	0.99	1.00
Wing length	3.63	3.71	3.83	···	5.40	5.57

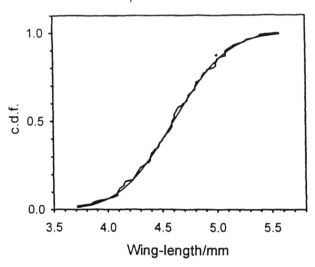

Figure 3.10 The c.d.f. obtained by plotting the cumulative probability in Table 3.2 against the wing length. The wiggly line gives the experimental values and the smooth line the theoretical c.d.f. for a normal distribution, with $m = 4.06$ mm and $s = 0.39$ mm.

We would also like to be able to calculate the theoretical c.d.f. corresponding to the theoretical p.d.f. given by Equation 3.9. Formally, the theoretical c.d.f. is the area under the curve below each value of x.[2] As it happens, the c.d.f. for the normal distribution cannot be obtained analytically, even using calculus, but we will see in Chapter 5 that the c.d.f. for the normal distribution is tabulated in books on statistics and is easily calculated with the aid of a computer. In Fig. 3.10 I have used a computer to calculate the theoretical c.d.f. for the housefly data and this is indicated by the smooth line.

Finally, we note that since

$$C(x_n) = \sum_{i=1}^{n} P(x_i)\delta x \qquad\qquad 3.14$$

and

$$C(x_{n-1}) = \sum_{i=1}^{n-1} P(x_i)\delta x, \qquad\qquad 3.15$$

$$C(x_n) - C(x_{n-1}) = P(x_n)\delta x, \qquad\qquad 3.16$$

so that we can calculate the p.d.f. simply by subtracting pairs of values from the c.d.f.[3]

3.3 SUMMARY

For both discrete and continuous random variables we can represent their distributions in two ways. The probability density function gives us the probability that a particular variable will fall into a given, small range. The cumulative distribution function gives the probability that a particular variable will be less than or equal to a particular value. At first sight the p.d.f. seems to be more fundamental and easier to grasp. However, as you will discover, we will be concerned almost exclusively with c.d.f.s and hardly at all with p.d.f.s. The reasons for this are threefold.

First of all, the p.d.f. is a density and to specify a particular value we need also to specify the units in which it is measured, whereas the c.d.f. is a pure probability and does not depend on the units in which it is measured. Secondly, if we know the c.d.f. we can immediately calculate the p.d.f. for any interval simply by subtracting the values of the c.d.f. at either end of the interval. Thirdly, as we will discover, when we examine our data we will be concerned to identify extreme events, ones that are unlikely to occur on a given hypothesis. The c.d.f. gives us directly the probability of particular extreme events occuring.

We do not always want to specify every distribution in all its detail. We therefore need to find ways to summarize the most important properties of our data in terms of a few key parameters. This is the subject of the next chapter.

NOTES

[1] For those who know calculus, the exact expression is

$$P(a \leqslant x \leqslant b) = \int_a^b P(x)dx.$$

[2] Using calculus

$$C(a) = \int_{-\infty}^a P(x)dx.$$

[3] For a continuous distribution

$$C(b) - C(a) = P(a \leqslant x \leqslant b) = \int_a^b P(x)dx$$

3.4 EXERCISES

1. (a) Using a computer, generate 600 evenly distributed discrete random numbers between 1 and 6 and plot the data as a bar chart. (b) Generate a second set of random numbers, add them to the first, and then plot the sum of the numbers as a histogram.

2. Estimate from Figs 3.3b and 3.4b the probability that the sum of the two numbers on a pair of dice shows 5 or less.

3. The data given in Table 3.3 and plotted in Fig 3.11 were published by the National Science Foundation in the United States of America. They show a decline during the 1960s and 1970s in the number of Nobel Prizes won by Americans after the numbers had peaked in the 1950s (Tufte, 1986). This graph was published to show that unless funding was increased, the United States of America would rapidly fall behind its competitors in scientific research. Why is this graph misleading? (Answer at the end of the examples.)

Representations

Table 3.3 The number of Nobel Prizes won by scientists in five countries from 1901 to 1974

Dates	USA	Germany	UK	USSR	France
1901–1910	1	12	5	2	6
1911–1920	2	7	3	0	5
1921–1930	4	8	7	0	4
1931–1940	9	8	7	0	3
1941–1950	14	5	7	0	0
1951–1960	29	3	9	5	0
1961–1970	26	6	12	4	6
1971–1974	13	1	7	0	0

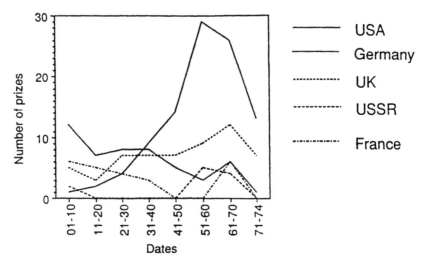

Figure 3.11 Nobel Prizes in science for selected countries, 1901–74.

4. Use Equation 3.9 and the values of m, s and N given in the text to calculate the number of flies whose wing lengths lie between 4.5 and 5.0 mm. Compare your result with the number given in Table 3.1. Use Fig. 3.10 to determine the number of flies whose wing lengths lie between 4.5 and 5.0 mm.

5. The following are the wing lengths of ten of the 100 flies given in Table 3.1, chosen at random: 4.40, 4.35, 4.25, 4.68, 5.40, 4.43, 3.63, 4.46, 4.43, 4.63. Plot the p.d.f. and the c.d.f. for these data and compare the plots with Figs 3.5 and 3.10.

[Answer to Question 3: Each point on the graph gives the number of prizes won in a 10 year period, except the last set of points which only cover 4 years. Not surprisingly, the number of prizes fell in all countries at the end of the plot. Plot the number of prizes won by scientists from each country **per decade** by multiplying the·numbers in the last row by 10/4.]

4

Measures

How can it be that mathematics, being after all a product of human thought which
is independent of experience, is so admirably appropriate to the objects of reality?
Is human reason, then, without experience, merely by taking thought, able to fathom
the properties of real things? In my opinion the answer to this question is briefly
this: as far as the propositions of mathematics refer to reality, they are not certain;
and as far as they are certain, they do not refer to reality.

A. Einstein (1954)

If you were asked to describe a giraffe, you might say that it is a very tall, four-legged
mammal with a long neck and a patchy brown coat. This description is brief; you
could elaborate on it at considerable length. But it is often useful to be able to
indicate one or two key features of an object that are sufficient to describe its form
or function. Of course, the particular features we name depend on our reasons for
wanting to classify the object. You would not give a child Linnaeus's description of
an elephant and you would not tell a biologist that an elephant is a large grey beast
with tusks. If you wished to distinguish a horse from a zebra you might say that
the zebra has stripes that the horse does not have, while to distinguish a horse from
a donkey, you might say that the donkey has long ears that the horse does not
have. In biology we constantly **classify** the objects in the world around us and we
will need to do the same in statistics. The 'objects' we are concerned with in statistics
are probability distributions. In the last chapter we discussed the uniform distribution
(Fig. 3.1), which we obtained when we threw one die, the triangular distribution,
which we found when we threw two dice and added the numbers together (Fig. 3.2),
and the normal distribution, obtained when we considered the length of the wings
of 100 houseflies (Fig. 3.7). In this chapter we will think about ways of classifying
probability distributions so that we do not need to specify the distributions in every
detail but rather can pick out the key properties as we need them.

4.1 MEASURES OF LOCATION

We will start by considering ways to determine the location of a distribution.
Figure 4.1 shows the even and triangular distributions from Figs 3.1 and 3.2 with
the two distributions plotted on the same scales.

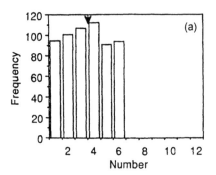
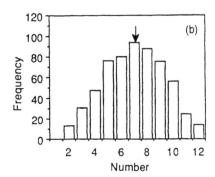

Figure 4.1 (a) The even distribution and (b) the triangular distribution shown in Figures 3.1 and 3.2. The arrows indicate the mean values.

4.1.1 Means

The first thing we might say about the two distributions is that the triangular distribution is shifted to larger values on the horizontal axis than is the even distribution. The most commonly used measure of location is the **mean** or average value. Using a horizontal bar to indicate means we have

$$\bar{x} = \sum_{i=1}^{n} x_i/n, \qquad\qquad 4.1$$

where n is the number of measurements that we make. For the even distribution x_i is the number that turns up on the ith throw; for the triangular distribution, x_i is the sum of the two numbers that turn up on the ith throw. In both cases n is the number of throws. For our housefly data, x_i is the length of the ith wing. We will also use $\langle \cdots \rangle$ to indicate average values or means since $\langle x \rangle$ is sometimes easier to write than a long bar over the top of an expression.

In Fig. 4.1 the mean of the numbers chosen from the even distribution is 3.48 and this is less than the mean of the numbers chosen from the triangular distribution which is 6.89. From Fig. 3.3 it is easy to see that the means of the theoretical even and triangular distributions are 3.5 and 7, respectively, so that our estimates from the simulation are quite close to the values for the underlying theoretical distributions. For the housefly data the mean wing length is 4.604 mm.

We use the Greek letter μ ('mu') for the 'true' mean of the underlying theoretical distribution to distinguish it from m, the mean estimated from the data in a particular experiment. For the even distribution $\mu = 3.5$, for the triangular distribution $\mu = 7$. (We will use Roman letters for parameters estimated from data and Greek letters for the equivalent parameters of the underlying distribution that we are trying to estimate.)

The mean can also be expressed in terms of the probability of each possible outcome. In the simulation in which we threw one die 600 times, we had 94 ones,

84 twos, 113 threes, 106 fours, 99 fives and 104 sixes, so that the mean m is

$$m = (1 \times 94 + 2 \times 84 + 3 \times 113 + 4 \times 106 + 5 \times 99 + 6 \times 104)/600$$

$$= 1 \times (94/600) + 2 \times (84/600) + 3 \times (113/600)$$

$$+ 4 \times (106/600) + 5 \times (99/600) + 6 \times (104/600)$$

$$= 1 \times P(1) + 2 \times P(2) + 3 \times P(3) + 4 \times P(4) + 5 \times P(5) + 6 \times P(6)$$

$$= \sum_{i=1}^{6} iP(i) = 3.5 \qquad\qquad 4.2$$

and in general the mean value of any variable x_i is the sum over x_i times the probability of getting x_i^1,

$$\bar{x}_i = \sum_i x_i P(x_i). \qquad\qquad 4.3$$

Just as we learned to combine probabilities, we need also to be able to combine means. Suppose, for example, we have calculated the mean length $\langle l \rangle$ of the left wings of 100 houseflies and the mean length $\langle r \rangle$ of the right wings of the same flies. If we now want $\langle l + r \rangle$, the mean length of all the wings, left and right, we can either add up the lengths of all the wings, left and right, and divide by 200, or simply note that provided we have the same number of wings in both cases, the mean of the sum is the sum of the means (as shown in the Appendix, section 4.5.1), so that

$$\langle l + r \rangle = \langle l \rangle + \langle r \rangle. \qquad\qquad 4.4$$

The mean has several especially desirable properties, but it is not the only measure of the location of a distribution and sometimes other measures are more appropriate. An alternative to the **arithmetic** mean given by Equation 4.1 is the **geometric mean**, G, which is the nth root of the product of the individual measurements, so that

$$G = (x_1 \times x_2 \cdots \times x_n)^{1/n}. \qquad\qquad 4.5$$

Taking logarithms of both sides of Equation 4.5 gives

$$\ln G = (\ln x_1 + \ln x_2 \cdots + \ln x_n)/n, \qquad\qquad 4.6$$

so that the logarithm of G is the arithmetic mean of the logarithms of the individual measurements.

4.1.2 Medians

Another measure of location is the **median**, which is the measurement that falls in the middle of the distribution so that there are as many items below it as above it. (For a continuous distribution, the median is the point that divides the area under the p.d.f. into two equal parts, which is also the point on the c.d.f. corresponding to a probability of 0.5.) For example, the mean of 1, 3, 4, 8 and 9 is 5, while the median is 4. For the housefly data given in Table 3.1, reading number 50 is 4.59 mm and reading number 51 is 4.60 mm, so we take 4.595 mm as the median wing length, which is close to the mean wing length of 4.604 mm.

As an example of a situation in which you might wish to use the median rather than the mean, suppose you carried out an experiment to measure the time it takes insects to die after they have been sprayed with an insecticide. If you start with 100 insects and then measure the time it takes each one to die, you can calculate the mean lifetime of the insects exposed to the insecticide. But if the last insect takes a long time to die, you might spend most of your time waiting for that one to die. To obtain the median lifetime, on the other hand, you need only to know the time at which the 50th insect dies, after which you can stop the experiment.

A further advantage of the median is that if we have outliers in our data, that is to say, a few points that are very much larger or very much smaller than the others, these points will influence the mean greatly. For example, the mean of 2, 4 and 6 is 4 and the mean of 2, 4 and 100 is 53, while in both cases the median is 4. There are many techniques that have been developed for handling outliers in data and these are discussed in detail by Winer (1971, p. 51).

If the distribution is symmetrical, the mean and the median of the underlying distribution are the same although the values calculated from the data may differ. One way to test if a set of numbers comes from a symmetrical distribution is to compare the mean and the median; if they differ significantly we conclude that the underlying distribution is not symmetrical.

4.1.3 Modes

Another measure of the location of a distribution is the **mode**, which is the location of the most probable outcome. For example, in our example of throwing two dice (Fig. 3.2), the mode estimated from the data is 7 and of course the mode of the theoretical probability distribution is 7. In the case of our housefly data as plotted in Fig. 3.6, the mode is 4.6.

4.2 MEASURES OF SCALE

Looking at the frequency histograms in Fig. 4.1, we see not only that the triangular distribution is shifted to the right of the even distribution, but the shapes of the two distributions differ; in particular, the triangular distribution is wider than the even distribution. So the next thing we need is a measure of the spread or width of our distributions.

4.2.1 Variance

To obtain a measure of the width of the distributions, we subtract the mean, since we have already taken that into account. For each measured point, we are then left with the deviation from the mean. One possible measure of the spread would be the average value of the deviations from the mean. However, the mean of 1, 2 and 3, for example, is 2 so that the deviations from the mean are -1, 0 and 1 and the mean deviation is 0! One way around this is to square the deviations before taking the average. For the three numbers given above the **squared deviations** are 1, 0 and 1 and the **mean squared deviation**, which we call the **variance**, is 2/3. The variance may be written

$$V = \sum_{i=1}^{n} (x - \bar{x})^2/n. \qquad 4.7$$

The variance is important in biological statistics where we are usually concerned to discover the causes of the variation in our data. In section 8.4.5 we will discuss an experiment to compare three different designs of tsetse fly traps. The number of flies that we catch depends on the design of the trap, the day of the experiment and the site in which each trap is placed. By designing the experiment carefully, it turns out that we can calculate a variance corresponding to each factor in our experiment. This technique, which is the subject of Chapter 8, is called the **analysis of variance**, or 'ANOVA' for short, and it tells us which factors (trap, day or site) contribute significantly to the variations that we observe in the data (number of flies caught).

Just as we learned to combine probabilities and means, we need to know the rules for combining variances. We can show (Appendix, section 4.5.2) that if we have two independent random variables, the variance of their sum is the sum of their variances, so that if, for example, $V(l)$ is the variance of the lengths of the left wings of 100 houseflies and $V(r)$ is the variance of the lengths of the right wings, the variance of the lengths of their right wings and their left wings, added together, is

$$V(l + r) = V(l) + V(r) \qquad 4.8$$

so that for independent random variables the variance, like the mean, is additive. If the left and right wing lengths are not independent we have an extra term in Equation 4.8 (see Equation 9.28).

We can also write an expression for the variance, as we did for the mean, in terms of probabilities:

$$V(x) = \langle (x - \bar{x})^2 \rangle = \Sigma (x - \bar{x})^2 P(x). \qquad 4.9$$

In fact, for any function of x we can always write the mean value of $f(x)$ as the sum over $f(x)$ times the probability of getting x, so that

$$\langle f(x) \rangle = \Sigma f(x) P(x). \qquad 4.10$$

Finally, we should note a computational formula that we will use in later chapters:

$$V(x) = \Sigma(x - \bar{x})^2/n = \Sigma x^2/n - 2\bar{x}\Sigma x/n + \bar{x}^2$$
$$= \Sigma x^2/n - (\Sigma x/n)^2 = \langle x^2 \rangle - \langle x \rangle^2. \qquad 4.11$$

We will use the right-hand side of Equation 4.11 to calculate variances in Chapter 8.

4.2.2 Standard deviation

Although ANOVA will be central to much of what we do, the units of the variance are not the same as those of our original data: when we measure the length of the wings of houseflies in mm, the variance has units of mm^2, and we would like to measure the width of the distribution in the same units as the original variable. For this reason we define the **standard deviation** as the square root of the variance (for our houseflies the units are then millimeters again) and

$$s = \left[\sum_{i=1}^{n} (x_i - \mu)^2/n \right]^{1/2} \qquad 4.12$$

The standard deviation is the root-mean-square value of the deviations about the mean.

As it stands, Equation 4.12 is correct. However, we do not usually know the true mean; rather, we have to use the value of the mean estimated from the data. Since the estimated mean minimizes the mean-square deviation about the mean, using m instead of μ in Equation 4.12 will systematically underestimate, or bias, the standard deviation. Fortunately, it is possible to show that if we divide by $n-1$ instead of n, this will remove the bias and our equation for estimating the standard deviation becomes

$$s = \left[\sum_{i=1}^{n} (x_i - m)^2/(n - 1) \right]^{1/2}. \qquad 4.13$$

This argument can be made more rigorous (Bulmer, 1979, p.130). What we are doing is dividing by the number of independent parameters we have in our estimate of s which we call the number of **degrees of freedom**. In our calculation of the standard deviation, we start with n data points, each of which is independent of the others, so that we have n degrees of freedom. When we calculate the mean we use up one of these degrees of freedom, leaving $n-1$, and this is the number we use in Equation 4.13.

For the even distribution, Fig. 4.1a, the standard deviation is 1.67 and for the triangular distribution, Fig. 4.1b, it is somewhat greater at 2.37. For the housefly data, the calculated standard deviation is 0.39 mm.

We use the Greek letter σ to indicate the 'true' value of the standard deviation so that the experimental standard deviation s is an estimate of σ.

4.2.3 Standard deviation of the mean

The mean value of a set of numbers is an estimate of the mean of the underlying distribution: so how accurate is this estimate of the underlying or 'true' mean?

Suppose we have a set of measurements x_i so that their mean is

$$m = \Sigma x_i/n. \qquad 4.14$$

Since the variance of a sum is the sum of the variances (Equation 4.8 and Appendix, section 4.5.2),

$$V(\Sigma x_i) = \Sigma V(x_i), \qquad 4.15$$

and if each x_i is chosen independently from the same distribution, they all have the same variance, s^2, so that

$$V(\Sigma x_i) = ns^2. \qquad 4.16$$

The standard deviation is the square root of the variance so that

$$s(\Sigma x_i) = \sqrt{ns}, \qquad 4.17$$

and the standard deviation of the **mean**, s_m, is

$$s_m = s(\Sigma x_i/n) = \sqrt{ns}/n = s/\sqrt{n}. \qquad 4.18$$

(Many authors call s the standard deviation and s_m the standard error. To avoid confusion, I call s the population standard deviation and s_m, the standard deviation of the mean.) The importance of Equation 4.18 is that it shows that as we make more measurements, the accuracy of each measurement, which is given by s, will remain constant but the accuracy of the mean, which is given by s_m, will improve as the number of measurements increases. Since the estimate of the mean improves as the square root of the number of measurements, we need to make four times as many measurements if we want to double the accuracy of our estimate.

4.2.4 Ranges

Just as the mean is not the only measure of the location of a distribution, the standard deviation is not the only measure of the dispersion. An alternative to the standard deviation as a measure of dispersion is the **range**, which is the difference between the largest and the smallest values in the data. For the even distribution (Fig. 4.1a), the range is 5, for the triangular distribution (Fig. 4.1b), it is 10 and for the housefly data (Table 3.1) it is 1.94 mm.

Both the range and the standard deviation, like the mean, are sensitive to the presence of outliers and so we sometimes use the **interquartile range**, which has properties rather like the median. The median is the point that occurs half-way up the distribution, and we define the **lower quartile** as the point that occurs one-quarter of the way up the distribution and the **upper quartile** as the point that occurs three-quarters of the way up the distribution. For example, for the numbers 1, 3, 4, 7 and 10, the lower quartile is 3, the upper quartile is 7 and the interquartile range is $7 - 3 = 4$. If the number 10 in the series is replaced by 100, the range and the standard deviation change considerably while the interquartile range remains the same.

4.3 MEASURES OF SHAPE

In nearly all of our work we will be concerned with means and variances or standard deviations, but we will sometimes want to know more about the shape of a distribution. We may want to know if a distribution is symmetrical or not. We may want to know if it has long tails extending out on either side or if it terminates abruptly. The two measures of the shape of a distribution that we will consider are the **skewness** and the **kurtosis**.

4.3.1 Skewness

The distributions we have considered so far are all symmetrical about the mean. As an example of a skew distribution that is not symmetrical about the mean, consider Table 4.1, which gives the number of Prussian army officers kicked to death each year by their horses in each of ten army corps in the 20 years from 1875 to 1894 (von Bortkewitsch, 1898).

From Table 4.1 it is clear that in most of the army corps in most years no officers were kicked to death. However, in 1882 four officers in corps 9 were kicked to death, and in three cells in Table 4.1 we see that three officers were kicked to death in the same corps in the same year. The most convenient way to represent the distribution of deaths is to calculate the frequency distribution so that we record the number of cells in Table 4.1 in which no officers were kicked to death, the number of cells in which one was kicked to death, and so on. The result is shown in Table 4.2.

The data in Table 4.2 are plotted as a bar chart in Fig. 4.2. The distribution is clearly **skew**. Since the tail is to the right, we say that the distribution is skew to the right; if the tail were on the other side, we would say that the distribution is skew to the left. We now need a measure of the skewness of the distribution shown in Fig. 4.2. We would like it to be zero for a symmetrical distribution, positive if the

Table 4.1 The number of Prussian army officers kicked to death by their horses in each of ten army corps in each year from 1875 to 1894

	Corps									
Year	1	2	3	4	5	6	7	8	9	10
75	.	.	.	.	1	1	.	.	1	.
76	.	.	1	.	.	.	.	.	1	1
77	.	.	.	.	1	.	.	1	2	.
78	2	1	1	.	.	.	.	1	1	.
79	.	1	1	2	.	1	.	.	1	.
80	2	1	1	1	.	.	2	1	3	.
81	.	2	1	.	1	.	1	.	.	.
82	.	.	.	.	.	1	1	2	4	1
83	1	2	.	1	1	.	1	.	.	.
84	1	.	.	.	1	.	.	2	1	1
85	.	.	.	.	.	.	2	.	.	1
86	.	.	1	1	.	.	1	.	3	.
87	2	1	.	.	2	1	1	.	2	.
88	1	.	.	1	.	.	.	.	1	.
89	1	1	.	1	.	.	1	2	.	2
90	.	2	.	1	2	.	2	1	2	2
91	.	1	1	1	1	1	.	3	1	.
92	2	.	1	1	.	1	1	.	1	.
93	.	.	.	1	2	.	.	1	.	.
94	.	.	.	.	.	1	.	1	.	.

Table 4.2 The frequency with which a given number of officers were kicked to death by their horses in ten Prussian army corps in each of the 20 years from 1875 to 1894. k_i deaths occurred in each of n_i of the cells in Table 4.1

Number of deaths, k_i	Observed frequency, n_i
0	109
1	65
2	22
3	3
4	1
5 +	0

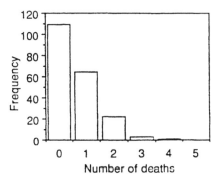

Figure 4.2 Bar chart of the data shown in Table 4.2.

distribution is skew to the right and negative if the distribution is skew to the left. We have already defined the mean as $\langle x \rangle$ and the variance as $\langle (x - \bar{x})^2 \rangle$, so it makes sense to define the skewness as $\langle (x - \bar{x})^3 \rangle$. But just as the variance of our housefly wing lengths has dimensions of mm², this definition of skewness would have units of mm³. We could take the cube root, which for our houseflies would have units of millimeters, but since the skewness is to be a measure of shape, we prefer it to be dimensionless, and so instead we divide by the cube of the standard deviation to get

$$S = \langle (x - \bar{x})^3 \rangle / s^3. \qquad 4.19$$

With this definition, a symmetrical distribution has skewness equal to zero. For example, the numbers -1, 0 and 1, for which the standard deviation is 1, have skewness $= \{[(-1)^3 + (0)^3 + (1)^3]/3\}/1^3 = 0$. Furthermore, the numbers -1, 0 and 4 have skewness 0.32 while the numbers -4, 0 and 1 have skewness -0.32. The numbers -2, 0 and 8 also have skewness 0.32, so that the skewness depends only on the shape of the distribution and not on the width, and it is positive when the tail is on the right, negative when the tail is on the left.

Returning to our Prussian army officers, the mean of the frequency distribution is

$$m = \Sigma k_i P(k_i) = \Sigma k_i n_i / n = 0.610/\text{corps/year} \qquad 4.20$$

where n_i is the number of cells in Table 4.1 in which k_i officers in any one corps were kicked to death in any one year, and n is the total number of cells in Table 4.1 which is $10 \times 20 = 200$. In 10 years, for example, we would expect an average of $0.61 \times 10 = 6$ army officers to be kicked to death in each corps. The variance is

$$V = \Sigma (k_i - m)^2 P(k_i) = \Sigma (k_i - m)^2 n_i / (n - 1) = 0.611/\text{corps}^2/\text{year}^2, \qquad 4.21$$

so that the standard deviation is 0.782/corps/year. The skewness is

$$S = \Sigma (k_i - m)^3 P(k_i)/s^3 = \{\Sigma(k_i - m)^3 n_i/n\}/s^3 = 1.24, \qquad 4.22$$

and as you can see from Figure 4.2, the distribution is skew to the right.

4.3.2 Kurtosis

The last measure we will consider is the **kurtosis**, K, which tells us if the distribution has long tails or short tails. If the distribution has long low tails and a narrow peak in the middle, the kurtosis is high and the distribution is called **leptokurtic**. If the distribution has a broad hump in the middle and short tails, it is called **platykurtic**. We define the kurtosis as the fourth power of the deviation about the mean divided by the fourth power of the standard deviation,

$$K = \langle (x - \bar{x})^4 \rangle / s^4, \qquad 4.23$$

so that it too is dimensionless.

Consider, for example, the two distributions in Table 4.3 illustrated in Fig. 4.3. Both distributions have a mean of 0, a standard deviation of 0.834 and, since they are symmetrical, a skewness of 0. However, they are clearly different and the difference is in the kurtosis, which is 4.2 for the first and 1.4 for the second. Note that the distribution with the longer tails has the greater kurtosis. We will see in the next chapter that the normal distribution has a kurtosis of 3 and this is taken as the standard value against which we compare the kurtosis of other distributions. For the

Table 4.3 Two frequency distributions. In the first the outcome i occurs m_i times. In the second the outcome i occurs n_i times. The mean, standard deviation and skewness are the same for both but m has a kurtosis 4.2 while for n it is 1.4

i	m_i	n_i
-2	1	0
-1	4	8
0	14	8
1	4	8
2	1	0

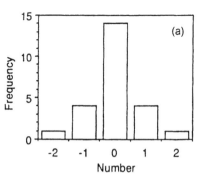
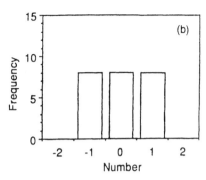

Figure 4.3 Bar charts of the distributions given in Table 4.3.

frequency distribution of ill-fated Prussian army officers, the kurtosis is

$$K = \{\Sigma(k_i - m)^4 n_i/n\}/s^4 = 4.37, \qquad 4.24$$

rather more than the standard value of 3. (Many statistics packages calculate Fisher's κ statistics rather than the skewness and the kurtosis and are discussed in the Appendix, section 4.5.3.)

4.4 SUMMARY

In this chapter we have developed ways of measuring the position, the width and the shape of statistical distributions. In most of what we do we will use the mean, the standard deviation or the variance, the skewness and the kurtosis. But we have also seen that there are other measures, such as the median and the interquartile range, that we could use instead, and if our data are not 'well behaved', for example if we have outliers that we are not sure of, these other measures may be more appropriate. As we will see in the next few chapters, the mean and the standard deviation enable us to make precise statements about our data but only if we know the 'true' underlying distribution. But suppose that we do not know what the underlying distribution is. Even worse, suppose that our data are not strictly quantifiable as would be the case if we only had categories such as very small, small, medium, large and very large. Provided we can order the categories, we can still determine measures such as the median while parameters such as the mean may be of little or no use. In these situations we will use what are called non-parametric statistics, of which the median is one, and we will discuss these further in Chapter 8.

4.5 APPENDIX

4.5.1 The mean of a sum

The mean of a sum is the sum of the means.

If we measure x_i and y_i for $i = 1, 2, \ldots n$ and let $z_i = x_i + y_i$, then $\langle z \rangle = \langle x \rangle + \langle y \rangle$.

We have

$$\langle z \rangle = \Sigma z_i/n = \Sigma(x_i + y_i)/n = \Sigma x_i/n + \Sigma y_i/n = \langle x \rangle + \langle y \rangle. \qquad 4.25$$

4.5.2 The variance of a sum or difference

For independent random variables, the variance of a sum or a difference is the sum of the variances.

Suppose that we measure two random variables, x and y, whose true means are μ and v. Then

$$V(x + y) = \sum_i \sum_j (x_i + y_j - \mu - v)^2 P(x_i \text{ and } y_j) \qquad 4.26$$

and if x and y are statistically independent,

$$P(x_i \text{ and } y_j) = P(x_i)P(y_j) \qquad 4.27$$

so that

$$V(x + y) = \sum_i (x_i - \mu)^2 P(x_i) \sum_j P(y_j) + \sum_j (y_j - v)^2 P(y_j) \sum_i P(x_i)$$

$$- 2\sum_i (x_i - \mu)P(x_i) \sum_j (y_j - v)P(y_j). \qquad 4.28$$

The last term is the mean value of $x - \mu$ times the mean value of $y - v$, both of which are zero. Furthermore,

$$\sum_i P(x_i) = \sum_j P(y_j) = 1, \qquad 4.29$$

so that

$$V(x + y) = V(x) + V(y). \qquad 4.30$$

If x and y are not independent, we have an extra term in the expression for the variance of the sum. We discuss this in section 9.5.

4.5.3 Fisher's κ statistics

Fisher's κ statistics are similar to the skewness and kurtosis described here but are adjusted for biases in small samples, the equivalent of dividing by $n - 1$ when we calculate the variance or the standard deviation (Bliss, 1967, p. 144).

Fisher's κ statistics S^* and K^* are

$$S^* = SN^2/[(N-1)(N-2)], \qquad 4.31$$

which, if the original data are normally distributed, is itself approximately normally distributed with mean zero and variance $6N(N-1)/[(N-2)(N+1)(N+3)]$ and

$$K^* = KN^2(N+1)/[(N-1)(N-2)(N-3)] - 3(N-1)^2/[(N-2)(N-3)], \quad 4.32$$

which, if the original data are normally distributed, is itself approximately normally distributed with mean zero and variance $4(N^2 - 1)V(S^*)/[(N-3)(N-5)]$. Fisher's κ statistics are used to test if the skewness and kurtosis of a given distribution differ significantly from the values obtained for a normal distribution.

NOTE

[1] For continuous functions

$$\langle x \rangle = \int_{-\infty}^{\infty} x P(x)\, dx$$

4.6 EXERCISES

1. Calculate the means, standard deviations and skewness of (a) $-1, 0, 1$; (b) $-1, 0, 4$; (c) $-4, 0, 1$; (d) $-2, 0, 8$.

2. Use Equations 4.3 and 4.10 to calculate the mean, standard deviation, skewness and kurtosis of (a) the even distribution generated by throwing one die and (b) the triangular distribution generated by throwing two dice and adding the numbers together.

3. Mark the position of the median of the data (4.595 mm) on the p.d.f. given in Fig. 3.5 and decide if it divides the area under the curve into two equal parts. Do the same for the c.d.f. given in Fig. 3.10 and decide if it corresponds to a probability of 0.5.

4. Calculate the standard deviation of the following numbers: 4.32, 2.79, 5.14, 3.87 and 2.31 using both $V(x) = \Sigma(x - \bar{x})^2/n$ and $V(x) = \langle x^2 \rangle - \langle x \rangle^2$. (These formulae use n, the number of measurements, rather than $n-1$, the number of degrees of freedom. In both cases we can multiply the final answer by $n/(n-1)$ to allow for this.

5. Determine the range, lower quartile, upper quartile and interquartile range for the housefly data given in Table 3.1

6. Use a computer to generate three sets of normally distributed random numbers with mean 0 and standard deviation 1, the first set containing ten numbers, the second 40 and the third 160. Calculate the mean, the population standard deviation and the standard deviation of the mean of each set of numbers. Note the relationships between these estimates for the three sets of numbers.

7. You measure the lengths of the wings of 16 flies and obtain a mean of 4.60 mm and a population standard deviation of 0.40 mm. What is the standard deviation of the mean? How many wings would you need to measure to obtain an estimate of the mean that was accurate to within 0.025 mm?

8. Work out the mean, standard deviation, skewness and kurtosis of the Prussian army officer data in Table 4.2.

5

Distributions

I know of scarcely anything so apt to impress the imagination as the wonderful form of cosmic order expressed by the 'Law of Frequency of Error'. The Law would have been personified by the Greeks and deified if they had known of it. It reigns with serenity and in complete self-effacement amidst the wildest confusion. The huger the mob, and the greater the apparent anarchy the more perfect is its sway. It is the perfect law of Unreason.   F. Galton (1889)

In the previous chapters we considered examples of random distributions: from the throw of a die we found the uniform distribution which is flat; from Mendel's experiments on peas we obtained a distribution that can take on only one of two values; from the lengths of the wings of houseflies we had the 'bell-shaped' normal distribution; and from the fate of Prussian army officers we found a skew distribution. In this chapter we will consider the most common distributions and their properties and we will discover that a small number of distributions will suffice to describe most biological situations.

We have already seen that if we throw two dice, the numbers that turn up on each one are uniformly distributed between 1 and 6, while the sum of the pairs of numbers on the two dice has a triangular distribution ranging from 2 to 12. We can therefore generate a triangular distribution by adding together pairs of numbers each taken from a uniform distribution. In this chapter we will examine several basic distributions as well as the relationships among them.

To convince you that we need to understand the properties of distributions, consider the data on the polio vaccine (Table 1.2). The vaccinated children appear to have gained some protection from polio, but few children succumbed to polio in either group and we need to know if the effect of the vaccine is statistically significant. In other words, does the difference represent a real effect of the vaccine or is it just chance that one number turned out to be less than the other? If we can establish the intrinsic variability of the data and if we can then show that the difference between the proportion of vaccinated and the proportion of unvaccinated children who develop polio is greater than this variability, we can reasonably argue that the vaccine did have a significant effect. But to determine the intrinsic variability of the data, we need to determine the underlying distribution, and this takes us into the subject of this chapter.

In this chapter we will discuss the **binomial, Poisson, normal,** χ^2 ('chi-square'), **t** and **F** distributions. We will start with the binomial distribution, which we get when there are only two possible outcomes, such as heads or tails, true or false, success or failure. When one of the two outcomes occurs only rarely, we will see that the binomial distribution may be approximated by the Poisson distribution. When the number of successes is sufficiently large, both the binomial and the Poisson distribution may be approximated by the normal distribution. Sometimes we will want to combine variables, each of which follows a normal distribution, and we will see that the sum of the squares of normally distributed variables follows a χ^2 distribution, while the ratio of a normally distributed variable to the square root of a variable that follows the χ^2 distribution follows a t distribution. Finally, we will see that the ratio of two χ^2 variates follows an F distribution.

5.1 THE BINOMIAL DISTRIBUTION

We get a **binomial** distribution when there are only two possible outcomes in an experiment, such as heads or tails, tall or short, round or wrinkled. Suppose, for example, that we want to know if a coin is biased. It is no good throwing it once since it will certainly land either heads or tails, but if we throw it 100 times and get 45 heads and 55 tails, we can then ask if this is or is not indicative of a bias in the coin. The binomial distribution will give us the probability of getting 45 heads and 55 tails, or indeed any other combination of heads and tails, and this will enable us to decide if the observed outcome is so unlikely that we should regard the coin as biased.

When there are only two possible outcomes, we use the words success and failure to describe the alternatives, and we write p for the probability of success and q for the probability of failure, bearing in mind that success and failure are arbitrary terms and may correspond to heads and tails, tails and heads, tall and short or any other pair of characteristics. Then since the probabilities of all possible outcomes must add up to 1 (section 2.2),

$$p + q = 1. \qquad 5.1$$

Now consider Table 5.1, which shows the various ways in which two or three coins can fall. If we throw two coins and apply our Law of Multiplication, the probability of getting one heads and one tails is pq and this can arise in two ways (HT or TH), so that using the Law of Addition, the probability of getting one heads and one tails in any order is $2pq$. In general the probability of getting k heads in n throws can be obtained by expanding $(p + q)^n$ and then choosing the kth term in the series. Alternatively, the probability of getting k successes in n trials may be calculated using (Bulmer, 1979, p. 84)

$$P(k \text{ successes}|n \text{ trials}) = \frac{n!}{k!(n-k)!}p^k q^{n-k} \qquad 5.2$$

where p is the probability of success and q is the probability of failure in any one

Table 5.1 The possible outcomes when throwing either two or three coins. The table gives the probability of each outcome, the number of ways the coins can land showing 0, 1, 2 or 3 heads, the relative frequencies of each combination of heads and the probability of getting each number of heads

			2 coins				
Outcome	HH		HT	TH		TT	
Probability	pp		pq	qp		qq	
Heads	2			1		0	
Frequency	1			2		1	
Probability	p^2			$2pq$		q^2	
			3 coins				
Outcome	HHH	HHT	HTH	THH	HTT	THT	TTH	TTT
Probability	ppp	ppq	pqp	qpp	pqq	qpq	qqp	qqq
Heads	3		2			1		0
Frequency	1		3			3		1
Probability	p^3		$3p^2q$			$3pq^2$		q^3

trial. (Remember that $n! = n \times (n-1) \times (n-2) \ldots 1$, $0! = 1$ and that $x^0 = 1$.) For example, the probability of getting two heads in three throws of a coin is

$$P(2|3) = \frac{3!}{2!\,1!}p^2 q^1 = 3p^2 q, \qquad 5.3$$

in agreement with the value given in Table 5.1. For an unbiased coin for which $p = q = 0.5$, we expect, on average, to get two heads in three throws with probability $P(2|3) = 3/8$.

To summarize: we get a **binomial distribution** whenever there are **only two possible outcomes, the probability of success or failure is constant from one trial to the next and each trial is independent of all the others.** Then if we know the probability of success in one trial, Equation 5.2 gives us the probability of getting k successes in n trials.

The binomial distribution is illustrated in Fig. 5.1. Keeping n, the number of trials, constant at 10 (Figs 5.1a, b and c), the distribution shifts to the right as p, the probability of success, increases from 0.1 to 0.9. When the probability of success is zero, the number of successes is always zero and when the probability of success is one, the number of successes is always equal to n. All three distributions peak at pn (1, 5 and 9 in Figs 5.1a, b and c, respectively) and we will see below that the mean of the binomial distribution is in fact pn. To see the effect of changing the number of trials, Fig. 5.1d shows the distribution for 100 trials with p set to 0.5. For $p = 0.5$, the width of the distribution for ten trials (Fig. 5.1b) is about 4, while the width of the distribution for 100 trials (Fig. 5.1d) is about 12, so that increasing the

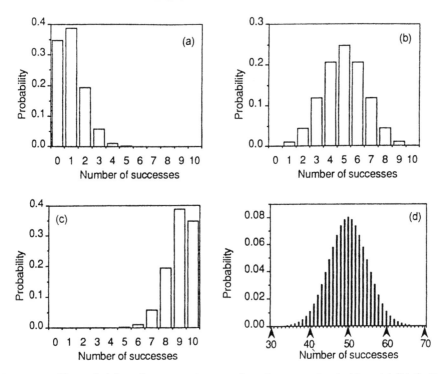

Figure 5.1 The probability of getting a given number of successes for the binomial distribution where the number of trials, n, and the probability of success, p, are (a) $n = 10$, $p = 0.1$; (b) $n = 10$, $p = 0.5$; (c) $n = 10$, $p = 0.9$; (d) $n = 100$, $p = 0.5$.

number of trials by ten times increases the width of the distribution by about three times. We will see below that the width is in fact proportional to the square root of the number of trials. The reason the distribution for 100 trials (Fig. 5.1d) looks narrower than the distribution for ten trials (Fig. 5.1b) is because of the change in the horizontal scale.

5.1.1 Families of eight

To illustrate the binomial distribution, consider the following problem. One occasionally finds large families in which all of the children are boys or all of the children are girls. Can these families of all boys or all girls be ascribed to chance alone or is there a genetic propensity for children in large families to be of the same sex? Table 5.2, extracted from birth registers in Saxony between 1876 and 1895 (Giessler, 1889), shows the number of boys and girls in families with eight children from a study of 53 680 families and we see, for example, that 215 families had 0 boys and 8 girls while 342 had 8 boys and 0 girls. We will treat each family as one experiment in which we have eight trials. For each trial there are two possible outcomes: a boy or a girl. We then repeat the experiment with each family in our sample.

Table 5.2 The number of families, f_i, having eight children of whom b_i are boys. p_i is the binomial probability of having b_i boys. F, the total number of families, is equal to 53 680 and Fp_i is the expected number of families of eight children with b_i boys

Number of boys, b_i	Number of families, f_i	Binomial probability, p_i	Expected number of families, Fp_i
0	215	q^8	165
1	1 485	$8pq^7$	1 401
2	5 331	$28p^2q^6$	5 202
3	10 649	$56p^3q^5$	11 034
4	14 959	$70p^4q^4$	14 627
5	11 929	$56p^5q^3$	12 411
6	6 678	$28p^6q^2$	6 581
7	2 092	$8p^7q$	1 994
8	342	p^8	264

In our binomial formula (Equation 5.2), n is equal to 8. For each sample we then have nine possible outcomes: $0, 1, \ldots, 8$ boys and $8, 7, \ldots, 0$ girls. To determine the expected numbers of families with $0, 1, 2, \ldots,$ boys we first need to know p, the probability of having a boy. The total number of boys is

$$B = 0 \times 215 + 1 \times 1485 \ldots 8 \times 342 = 221\,023, \qquad 5.4$$

and in the same way we find that the total number of girls is

$$G = 208\,417. \qquad 5.5$$

The proportion of boys, which gives an estimate of p, the probability of having a boy, is then

$$p = 221\,023/(221\,023 + 208\,417) = 0.5147, \qquad 5.6$$

so that q, the probability of having a girl, is

$$q = 1 - 0.5147 = 0.4853. \qquad 5.7$$

The first thing we notice is that the probability of having a boy is slightly greater than the probability of having a girl. But is this just due to chance or is there a genetic propensity to produce slightly more boys than girls? We can work out the various parameters of the distribution. Let us write b_i for the number of boys in a family (column 1 in Table 5.2) and f_i for the number of families (column 2 in Table 5.2) with b_i boys. Then, the total number of families is

$$F = \sum_i f_i = 53\,680. \qquad 5.8$$

(As a check, eight times the number of families must give the total number of children: $8 \times 53\,680 = 429\,440 = 221\,023 + 208\,417$.) The mean number of boys is then (using Equation 4.2)

$$m = \sum_i b_i p_i = \sum_i b_i f_i / F = 4.118. \qquad 5.9$$

The variance of the number of boys is (using Equation 4.9)

$$V = \sum_i (b_i - m)^2 f_i / (F - 1) = 2.067,$$ 5.10

so that the population standard deviation of the number of boys is $\sqrt{2.067} = 1.438$. The standard deviation of the mean number of boys is (Equation 4.18)

$$s_m = s / \sqrt{F} = 0.00621.$$ 5.11

Let us now see if the slight preponderance of boys is significant or if it is simply a result of the inherent variation in the data. If the probability of having a boy was equal to the probability of having a girl, the mean number of boys per family should be 4 whereas the observed value is 4.118, so that the observed and expected values differ by 0.118. Since the standard deviation of the mean is 0.0062, the observed and expected values differ by 19 times the standard deviation, the amount of spread we expect, and the difference is almost certainly real and not a matter of chance.

We can now consider the question that we began with: does the fact that 557 families have all boys or all girls indicate that some people have a propensity to produce children of the same sex or can it be put down to the fact that with so many families in our sample we would expect about this many to have all boys or all girls?

To do this we use the binomial probabilities and the values we have calculated for p and q to work out how many families we expect to have $0, 1, 2 \ldots 8$ boys (assuming that the assumptions of the binomial distribution hold), as shown in column 4 of Table 5.2. The agreement is reasonably good although the observed numbers of families are about 30% higher than the expected numbers at the ends of the distribution. If there is a genetic propensity to produce all boys or all girls, it must be rather slight and we still need to decide how big the difference should be before we consider that it is significant.

For the binomial distribution, it is possible to derive expressions for each of the parameters in terms of p, q and n, the number of trials. The mean, μ, is

$$\mu = pn.$$ 5.12

Since we don't know the 'true' probability of having a boy, we have to use our estimated probability of having a boy (Equation 5.6) to estimate the mean and we have $0.5147 \times 8 = 4.118$, the same value as before (Equation 5.9).

The variance is more interesting; we can show that for a binomial distribution the variance may be written in terms of p, q, and n as (Bulmer, 1979, p. 90)

$$V = pqn,$$ 5.13

which for our families of eight children gives $4.117 \times 0.4853 = 1.998$, very close to the value of 2.067 estimated directly from the data using the squared deviations from the mean.

We can do the same for the skewness and the kurtosis. We can calculate the skewness directly from the data as

$$S = \left\{ \sum_i (b_i - m)^3 f_i / F \right\} / V^{3/2} = -0.016,$$

5.14

and the kurtosis as

$$K = \left\{ \sum_i (b_i - m)^4 f_i / F \right\} / V^2 = 2.787.$$

5.15

But we can also show that for a binomial distribution the skewness is (Bulmer, 1979, p. 90)

$$S = (q - p)/\sqrt{npq},$$

5.16

which for our families of eight children gives -0.021, and the kurtosis is (Bulmer, 1979, p. 90)

$$K = 3 + (1 - 6pq)/npq$$

5.17

which for our families of eight children gives 2.75. We see that the variance, skewness and kurtosis estimated from p, q, and n agree well with the values estimated directly from the data, indicating that the data are close to a binomial distribution. The expressions for the mean, variance, skewness and kurtosis of the binomial distribution are summarized in Table 5.3.

We see that we are able to estimate the parameters of the distribution in two ways: either directly from the data using the probability that a family has 0, 1, 2...8 boys, or by using the data to estimate p, the probability of having a boy, from which q follows immediately, and n, the number of children in each family, in this case 8. For a given number of trials, the binomial distribution has only one free parameter, p, since q is always $1 - p$, and we use this in calculating the mean. However, the expressions in Table 5.3 enable us to calculate the other parameters once we know p and n and if the underlying distribution is indeed binomial we should get very similar values whichever way we calculate these parameters.

We can also use the expressions given in Table 5.3 to estimate the standard deviations of the numbers of children in each category separately to help us to decide if the differences between the observed and expected numbers are significant or not. To do this, let us look at the problem in a different way and consider each

Table 5.3 The mean, variance, skewness and kurtosis for the binomial distribution in terms of the number of trials, n, the probability of success, p, and the probability of failure, q

$$m = pn$$
$$V = pqn$$
$$S = (q - p)/\sqrt{npq}$$
$$K = 3 + (1 - 6pq)/npq$$

family as a trial (rather than each birth in each family) and consider only two outcomes: 0 boys and 8 girls will be a success, all other combinations will be a failure. The number of trials is now equal to the number of families, which is 53 680, and the number of successes is the number of families having 0 boys and 8 girls. Then

$$p = 215/53\,680 = 0.0040, \qquad\qquad 5.18$$

$$q = 1 - 0.00401 = 0.9960, \qquad\qquad 5.19$$

$$n = F = 53\,680. \qquad\qquad 5.20$$

Now we have only a single repeat of this 'experiment' so we cannot deduce the variance, skewness or kurtosis directly from the data. However, the expressions given in Table 5.3 still apply, so that the variance is

$$V = pqn = 214, \qquad\qquad 5.21$$

and the standard deviation is

$$s = \sqrt{V} = 14.6. \qquad\qquad 5.22$$

Since the observed number of families is 215 and the expected number of families is 165 (Table 5.2), the observed and expected numbers differ by 50, which is $50/14.6 = 3.4$ standard deviations, so that it appears as though the difference might be significant. We could of course repeat this calculation for each combination of boys and girls and hope to have a more sensitive test. We will do this in section 6.2.

Note also that the variance of the number of families having eight boys, 214, is very close to the number of families having eight boys, 215, which is our estimate of the mean number of families having eight boys. Looking at Table 5.3 we see immediately that since, in this case, p is very small, q is very close to 1 and the mean and the variance are both equal to pn.

5.1.2 Mendel's peas

As another application of the binomial distribution, consider again Mendel's study on peas. In one experiment (Table 2.2), he examined 1064 plants in the F_2 generation, of which 787 were tall and 277 were short. According to Mendel's theory, the probability of obtaining a tall plant is 0.75, and there are only two possible outcomes, tall and short. If we assume that the events are independent, so that what happens to one plant does not influence what happens to another, the distribution of plants should follow a binomial distribution. We could, of course, work out the precise probability of obtaining 787 tall plants from a total of 1064 plants with a probability of 0.75 that each plant is tall, but most of the information we need is contained in the mean and the variance. Since the mean of a binomial distribution is pn, the expected value of the mean is $1064 \times 0.75 = 798$, so that the difference between the observed and expected number of tall plants is $798 - 787 = 11$. Furthermore, the variance is $pqn = 0.75 \times 0.25 \times 1064 = 199.5$, so that the standard deviation is $\sqrt{199.5} = 14$. We see that the difference between the observed and expected number of tall plants is less than the expected spread in the number of tall plants, and we can conclude that Mendel's data agree with the predictions of the theory within the intrinsic variation of the experiment.

5.2 THE POISSON DISTRIBUTION

When the number of trials is relatively small the binomial distribution is easily calculated, but for the study of families with eight children, calculating the probabilities for each possible outcome was quite tedious. In an experiment in which the number of trials runs to hundreds or even thousands, it is impractical even to think about calculating the binomial frequencies exactly. However, it turns out that if the probability of success in each trial is very small but the number of trials is very large, the binomial distribution tends to a simpler form discovered in 1837 by the French mathematician Simeon-Denis Poisson.

For the binomial distribution, the expected number of successes in n trials is pn where p is the probability of success. Now suppose that we have a very large number of trials so that $n \to \infty$. (We read '$\to$' as 'tends to'. Since n cannot actually be infinite the idea is that $n \to \infty$ means that n is very much bigger than anything else in the problem. Similarly, $p \to 0$ means that p is very much smaller than anything else in the problem.) To ensure that the number of successes remains finite, we must have a very small probability of success so that we let $p \to 0$ in such a way that pn remains finite. These conditions define the **Poisson distribution**, which we get whenever **the number of trials is very large but the probability of success on each trial is very small**.

Whereas the binomial distribution is specified by two parameters, the number of trials, n, and the probability of success on each trial, p, the Poisson distribution is specified by a single parameter, $\mu = pn$, the mean number of successes. The probability of k successes is (Bulmer, 1979, p. 90)

$$P(k) = e^{-\mu}\mu^k/k! \quad k = 0, 1, 2 \cdots, \qquad 5.23$$

so that provided we know the mean number of successes, we can calculate the probability of getting any given number of successes.

Figure 5.2 shows the Poisson distribution for values of the mean, μ, equal to 1, 5, 9 and 50. The distribution in Fig. 5.2a is similar to the binomial distribution shown in Fig. 5.1a because q is small (0.1) and in both cases the mean is 1. As the mean increases, the distribution shifts to higher values and the width of the distribution increases as the square root of the mean while the ratio of the width to the mean decreases as the square root of the mean.

5.2.1 Prussian army officers

The horses in the Prussian army (von Bortkewitsch, 1898), which we discussed in section 4.3.1, must have had very many occasions on which they might have kicked their officers to death so that the number of trials in that experiment was indeed very large and we can let $n \to \infty$. The probability that any one officer was kicked

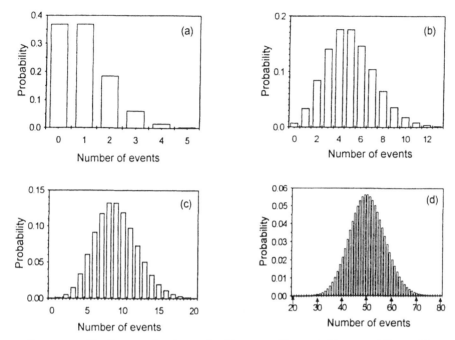

Figure 5.2 The Poisson distribution for (a) $m = 1$; (b) $m = 5$; (c) $m = 9$; (d) $m = 50$.

to death each time he happened to walk behind his horse was, fortunately for the Prussian army, very small and we can let $p \to 0$.

We don't know the actual number of trials, which would be the number of times an officer in each corps walked behind his horse in any one year, but for the Poisson distribution we need to know only the mean number of successes. In section 4.3.1 we showed that, m, the mean number of deaths per army corps per year, was 0.61. Assuming that the data follow a Poisson distribution, the probability that no deaths occur in any one corps in any one year is (Equation 5.23)

$$P(0) = e^{-0.61} 0.61^0/0! = e^{-0.61} = 0.5435. \qquad 5.24$$

Similarly, the probability that, say, three deaths occur in any one corps in any one year is

$$P(3) = e^{-0.61} 0.61^3/3! = 0.0205. \qquad 5.25$$

To get the **frequency** with which we expect no deaths, one death, and so on, we need to multiply each probability by 200, the total number of events (20 years × 10 corps). Table 5.4 gives the observed and expected frequencies of deaths and the two distributions agree very well indeed.

Since the Poisson distribution is the limiting form of the binomial distribution as $p \to 0$ and $q \to 1$, we can use the binomial results (Table 5.3) to show that for the Poisson distribution the variance is equal to the mean, μ, the skewness is equal to $1/\sqrt{\mu}$ and the kurtosis is equal to $3 + 1/\mu$, as summarized in Table 5.5

Table 5.4 The frequency with which between 0 and 5 officers were kicked to death by their horses in each year from 1875 to 1894 in each of ten Prussian army corps

Number of deaths	Observed frequency	Expected frequency
0	109	108.7
1	65	66.3
2	22	20.2
3	3	4.1
4	1	0.6
5	0	0.1

Table 5.5 The variance, skewness and kurtosis for the Poisson distribution

$$V = \mu$$
$$S = 1/\sqrt{\mu}$$
$$K = 3 + 1/\mu$$

We have already calculated the parameters of the distribution for the Prussian army data. The mean was found to be 0.61 deaths/corps/year, so that on average rather less than one officer was kicked to death by his horse in each corps in each year. The variance we estimated from the data as 0.611 death^2/corps^2/year^2, so that the variance is approximately equal to the mean as expected for a Poisson distribution. The skewness, estimated directly from the data, was 1.26, a little less than 1.28, the value calculated from the mean, and the kurtosis estimated directly from the data was 4.38, a little less than 4.64, the value calculated from the mean. The agreement between the values of the variance, skewness and kurtosis calculated directly from the data and from the expressions in Table 5.5 give us reason to believe that the data do follow a Poisson distribution.

5.2.2 The polio vaccine

Let us consider the data on the polio vaccine given in section 1.1.3. Of the 200 745 children vaccinated, 33 developed the disease. Since very many children were inoculated, n is very large, but since few developed the disease, p is very small. We therefore assume that the data follow a Poisson distribution and that the variance is equal to the mean number of children who develop the disease. Since 33 vaccinated children developed polio, the standard deviation is $\sqrt{33} = 5.7$. Similarly, 115 unvaccinated children developed the disease so that the standard deviation in this number is $\sqrt{115} = 10.7$.

Now the number of children in the two categories are not the same but if we calculate the number per 100 000 who contracted the disease we get 16.4 ± 2.86 for those who were vaccinated and 57.1 ± 5.33 for those who were not vaccinated. The difference between the two means, 40.7, is much greater than the expected spread in either measurement, 2.86 and 5.33, and we conclude that the vaccine does indeed bring about a significant degree of protection. If we want to be more precise we can apply our formula for the variance of a difference (Appendix, section 4.5.2), which gives $2.86^2 + 5.33^2 = 36.59$, so that the standard deviation of the difference is $\sqrt{36.59} = 6.04$ and the observed difference is about seven times the standard deviation.

We can now see what would have happened if we had used only one tenth as many children in the trial. We would then have vaccinated 20 000 children and we would expect to get about 33/10 or 3.3 cases of polio. Placebo would have been given to 20 000 children and we would expect to get 115/10 or 11.5 cases of polio. The expected value of the difference would then be 8.2, but the variance of the difference would be $3.3 + 11.5 = 14.8$, giving a standard deviation of $\sqrt{14.8} = 3.9$, so that the observed difference is only twice the standard deviation. The intrinsic variation in the data may well have concealed the difference between the vaccinated and the unvaccinated children.

The fact that the variance of a Poisson distribution is equal to the mean is important in the theory of sampling because it implies that for random sampling from large populations, for which the probability that any particular individual is trapped or caught is small, the standard deviation of the number sampled is given by the square root of the number sampled.

5.3 THE NORMAL DISTRIBUTION

The most important distribution in biology, if not in all of statistics, is the normal distribution. We have already touched on it in section 3.2.2, where we discussed the length of the wings of houseflies. The normal distribution was first investigated by Karl Gauss in 1809 in his study of the theory of errors in astronomical investigations (Bulmer, 1979, p. 108). The theory of errors is dominated by the normal distribution because the sum of any large number of random variables is normally distributed, a result known as the **Central Limit Theorem** (Bulmer, 1979, p. 115). There are, as always, some provisos: the contributing variables should be independent of each other and their variances should be well-behaved. Since it is often the case in nature that variables, such as the height of a person, are the result of many random effects, each of which contributes a little to the total, we often find that biological observables are normally distributed.

We have already seen (Equations 3.9 and 3.10) that the probability density function, or p.d.f., for a normal distribution with mean μ and standard deviation σ is (Bulmer, 1979, p 108)

$$P(x) = \frac{1}{\sigma\sqrt{2\pi}} e^{-(x-\mu)^2/2\sigma^2}. \qquad 5.26$$

The normal distribution is plotted in Fig. 5.3 for various values of the mean and standard deviation. As the mean μ, increases, the distribution shifts to the right and

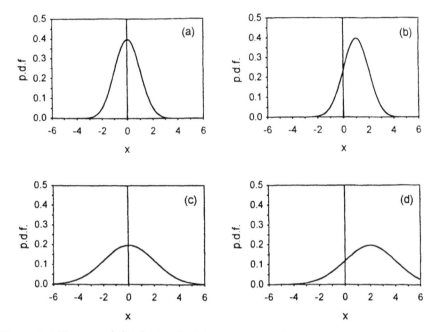

Figure 5.3 The normal distribution for (a) $\mu = 0$, $\sigma = 1$; (b) $\mu = 1$, $\sigma = 1$; (c) $\mu = 0$, $\sigma = 2$; (d) $\mu = 2$, $\sigma = 2$. The distribution is symmetrical about the mean, μ, and the width is proportional to the standard deviation, σ.

as the standard deviation, σ, increases, the distribution becomes wider. The normalization constant, $1/\sigma\sqrt{2\pi}$, is chosen to ensure that the area under the p.d.f. from $-\infty$ to ∞ is 1.

The normal distribution is symmetrical about the mean since we obtain the same answer if we set x equal to $\mu + \delta$ or $\mu - \delta$ in Equation 5.26; if we set $\mu = 0$, it will be centred at the origin. The standard deviation determines the width of the distribution and when σ is 1 and the mean is zero we refer to the **standard normal distribution**, indicated by the letter Z, so that

$$Z(x) = \frac{1}{\sqrt{2\pi}} e^{-x^2/2} \qquad 5.27$$

The standard normal distribution is tabulated in many books on statistics. We can use it to calculate $P(x)$ for any normal distribution with mean μ and variance σ^2 by noting that $P(x) = (1/\sigma)Z[(x - \mu)/\sigma]$. Since the normal distribution is symmetrical, its skewness is zero. The kurtosis of the normal distribution is 3 and we take this for the standard value of kurtosis.

5.3.1 Normal approximations

The Central Limit Theorem tells us that the sum of a large number of independently distributed random numbers is nearly always normally distributed provided the

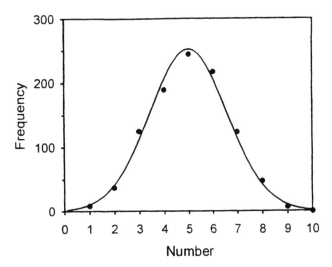

Figure 5.4 The distribution of numbers obtained when a coin was thrown 10 000 times and the numbers were added together in groups of ten counting 1 for heads and 0 for tails. The curve is the normal distribution that best fits the data.

individual distributions satisfy some reasonable conditions (Bulmer, 1979, p. 115). To illustrate this, I spun a coin (simulated on a computer) 10 000 times, counting 1 for heads and 0 for tails, and then added the numbers together in groups of ten. Figure 5.4 shows the distribution of the resulting numbers, with the normal distribution that best fits the data, and the agreement is good.

The mean and the standard deviation, calculated from the simulation, were 4.95 and 1.57. Using Equation 4.3, the mean for one throw of the coin is

$$\mu = 1 \times P(H) + 0 \times P(T) = 0.5, \qquad 5.28$$

and since the mean of a sum is the sum of the means, the expected mean for ten throws is 5, so that the value from the simulation (4.95) is close to the expected value. Using Equation 4.9, the variance for one throw of the coin is

$$V = (1 - 0.5)^2 \times P(H) + (0 - 0.5)^2 \times P(T) = 0.25, \qquad 5.29$$

and since the throws are independent, the variance of ten throws is 2.5 and the standard deviation is $\sqrt{2.5} = 1.58$, so that the value from the simulation (1.57) is again close to the expected value. The skewness and the kurtosis, calculated directly from the data, are -0.0062 and 2.909, close to the expected values of 0 and 3.

5.3.2 Binomial and Poisson distributions

Whenever there are two possible outcomes, success and failure, and the probability of success or failure is constant and independent from one trial to the next, the binomial distribution gives us the distribution of successes for any given number of trials. Furthermore, if the number of trials is very large and the probability of success on each is very small, the binomial distribution may be approximated by the Poisson distribution.

Both the binomial and the Poisson distributions are discrete so that they give us the probability of observing 0, 1, 2,...events. The normal distribution, on the other hand, is continuous and the observable may take on a range of continuous values, as in the case of the wing lengths of our houseflies. However, it is possible to show that provided the variance of the distribution is not too small, the binomial and Poisson distributions can both be approximated by a normal distribution. Referring back to Figs 5.1d and 5.2d, both the binomial and the Poisson distribution look like the normal distribution when the expected number of successes and the variance are sufficiently large. In general, the approximation is quite good provided the variance, which is pqn for the binomial and is equal to the mean for the Poisson distribution, is greater than about 9.

For the Prussian army data of Table 5.4, $pqn = 0.61$, and the distribution is quite skew, so we would not expect the distribution to approximate closely to a normal distribution. For our data on the numbers of boys and girls in families of eight in Saxony, we have $pqn = 2$, but the probability of success and failure are both close to 0.5 so that the distribution is almost symmetrical and the skewness is small (-0.02). Furthermore, the kurtosis (2.8) is close to 3, so let us see how well we can approximate the data in Table 5.2 using a normal distribution.

Equation 5.26 gives the p.d.f. for the normal distribution. Since the normalization constant is chosen so that the total area under the curve is equal to 1, we must multiply by 53 680, the total number of families, to obtain the frequency distribution $N(x)$. We therefore evaluate

$$N(x) = \frac{N}{s\sqrt{2\pi}}e^{-(x-m)^2/2s^2} \qquad\qquad 5.30$$

Table 5.6 The observed and expected numbers of families of eight children with between 0 and 8 boys. The third column gives the expected numbers after fitting the data to a binomial distribution and the fourth column the expected numbers after fitting the data to a normal distribution

No. of boys	Observed no. of families	Expected Binomial	Expected Normal
0	215	165	247
1	1 485	1 401	1 420
2	5 331	5 202	5 038
3	10 649	11 034	11 015
4	14 959	14 627	14 846
5	11 929	12 411	12 335
6	6 678	6 581	6 318
7	2 092	1 994	1 995
8	342	264	388

with a mean, m, of 4.118 and a standard deviation, s, of 1.438. The results are given in Table 5.6, from which we see that the normal approximation is too high in the tails (the binomial is too low), but the fit over the peaks is somewhat better and the overall fit is about as good as that obtained using the binomial distribution. The difference between the observed numbers and the numbers predicted using the binomial distribution may be due in part to random effects in the way the data were collected and recorded, which would tend to make the distribution look more like a normal distribution.

5.4 THE χ^2, t AND F DISTRIBUTIONS

There are three more distributions that are widely used in biostatistics and all may be derived from the normal distribution. In this section I will introduce them briefly and we will consider their properties in more detail as we use them.

5.4.1 The χ^2 distribution

Much of our work will be concerned with analysing the variances of the numbers we measure. When we test a new trap to see how well it catches tsetse flies, we always expect some variation in the number of flies caught. We will then want to know if the variation is simply due to the stochastic nature of the sampling process or if it is due to a variable of interest such as the colour of the trap, the site in which it is placed or even the day on which it is used. We have already seen that when we calculate the variance, we add up the squares of the deviations of each measurement about the mean. We generally assume that each of our measurements is normally distributed about the mean, and we would like to know how the sum of the squares of such normally distributed numbers will itself vary so that we can determine the distribution of our variances.

Suppose we let x be the sum of the squares of a number of independent, standard, normal variates, z_i, so that

$$x = z_1^2 + z_2^2 + \cdots + z_f^2 \qquad 5.31$$

Then x is said to belong to a **chi-square** distribution with f degrees of freedom, which is written χ_f^2. Figure 5.5 shows the χ^2 distribution for 1, 4, 16 and 64 degrees of freedom. You see that the distribution starts off being very skew but looks more like a normal distribution as the number of degrees of freedom increases. The mean of the χ^2 distribution is equal to the number of degrees of freedom, so that the distribution shifts to higher values as the number of degrees of freedom increases.

The parameters of the χ^2 distribution are given in Table 5.7, from which it is clear that as the number of degrees of freedom increases, the skewness tends to 0 and the kurtosis tends to the normal value of 3, so that the χ^2 distribution tends to the normal distribution. Of course, this is simply another example of the Central Limit Theorem, since we have defined the χ^2 distribution as the sum of a number of

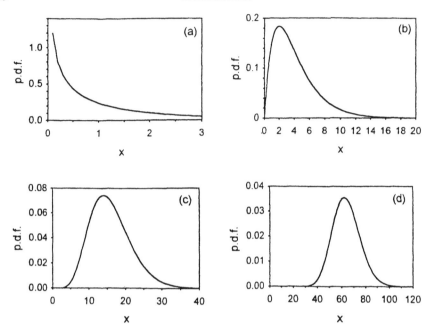

Figure 5.5 The χ^2 distribution for (a) 1, (b) 4, (c) 16 and (d) 64 degrees of freedom.

Table 5.7 Parameters of the χ^2 distribution

mean	f
variance	$2f$
skewness	$\sqrt{8/f}$
kurtosis	$3 + 12/f$

independent random variables and if we have enough of them their sum must follow a normal distribution.

5.4.2 The *t* distribution

We often want to compare the mean value of a measurement with its standard deviation: if the mean is less than the standard deviation, it is clear that it does not differ significantly from zero, while if it is much greater than the standard deviation, it is clear that it does differ significantly from zero. For example, in section 5.1.2, in our examination of Mendel's data on peas, we compared the difference between the observed and expected number of tall plants (11) with the standard deviation in the observed number (14). Because the difference was less than the standard deviation, we concluded that the difference was not significant and that the data supported Mendel's hypothesis. In section 5.2.2 we found that the probability of a child developing polio was reduced if the child had received the Salk vaccine. To convince ourselves that the effect was significant, we calculated the difference between the proportion of vaccinated and the proportion of unvaccinated children who developed polio (40.7 per 100 000) and then calculated the standard deviation of the difference (6.04 per 100 000). Because the difference was seven times the standard deviation, we concluded that the difference was real and the vaccine efficacious.

Let us restate this: if the ratio of the absolute value of the mean to the standard deviation is much less than 1, we know that the mean does not differ significantly from zero while, if the ratio is much greater than 1, we know that the mean does differ significantly from zero. If we can determine the distribution followed by the ratio of the mean to the standard deviation, then, as we will see in Chapter 7, we will be able to put precise limits on such statements and determine the probability that our observed value of the mean is simply due to chance as opposed to a real effect.

The distribution of the ratio of the mean to its standard deviation was first investigated by William Gossett (Student, 1908) who wrote under the pen name 'Student'. Since the mean is normally distributed and the variance follows a χ^2 distribution, we are interested in the ratio of a normally distributed variable to the square root of a variable following a χ^2 distribution; this is called **Student's t** distribution. Formally, if

$$x = z/\sqrt{y/f} \qquad 5.32$$

and z follows a normal distribution while y follows a χ^2 distribution with f degrees of freedom, we say that x follows a t distribution with f degrees of freedom. At this point we will extend our notation further and use the symbol $\sim$ to mean 'follows'. Then we can summarize this result as follows:

$$\text{If } z \sim N(0, 1) \text{ and } y \sim \chi_f^2,$$

$$\text{then } x = z/\sqrt{y/f} \sim t_f \qquad 5.33$$

Table 5.8 gives the parameters of Student's t distribution, from which we see that the distribution is symmetrical about the origin and that the kurtosis tends to the normal value of 3 as the number of degrees of freedom increases. Plotting the t distribution for various numbers of degrees of freedom (Fig. 5.6), we see that the t distribution looks very much like the normal distribution except that it has long tails, which become less pronounced as the number of degrees of freedom increases.

Table 5.8 Parameters of the t distribution. f is the number of degrees of freedom. For $f \leqslant 2$ the variance is infinite and for $f \leqslant 4$ the kurtosis of infinite

mean	0	
variance	$f/(f-2)$	$f > 2$
skewness	0	
kurtosis	$3 + 6/(f-4)$	$f > 4$

Distributions

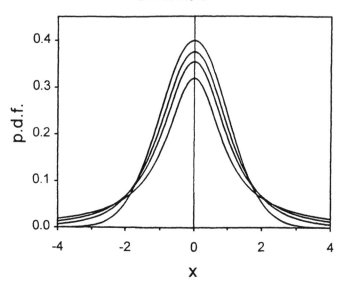

Figure 5.6 The *t* distribution for 1, 2, 4 and 100 degrees of freedom. As the number of degrees of freedom increases, the peak value increases and the tails become shorter.

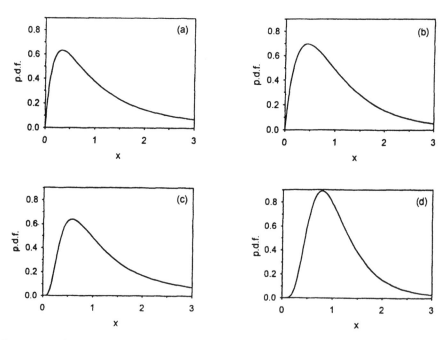

Figure 5.7 The *F* distribution with (a) 4 and 4; (b) 4 and 16; (c) 16 and 4 and (d) 16 and 16 degrees of freedom.

Table 5.9 The mean and the variance of the F distribution for large numbers of degrees of freedom. The expressions for the skewness and kurtosis are more complicated

mean	1
variance	$2(f_1 + f_2)/(f_1 f_2)$

5.4.3 The F distribution

We will use the t distribution to test a mean or to compare two means. Sometimes, however, we will want to compare two variances. To see if two variances differ significantly, we can calculate their ratio and then see if this ratio is greater or less than 1. Since we know that a variance follows a χ^2 distribution, we will want to know the distribution of the ratio of two numbers, each of which follows a χ^2 distribution; this is called the F distribution, after the English statistician Ronald Fisher. Then if

$$x = (y_1/f_1)/(y_2/f_2) \qquad 5.34$$

and y_1 and y_2 independently follow χ^2 distribution with f_1 and f_2 degrees of freedom, respectively, x follows an F distribution with f_1 and f_2 degrees of freedom. We can summarize this as follows.

$$\text{If } y_1 \sim \chi^2_{f_1} \text{ and } y_2 \sim \chi^2_{f_2} \text{ then } x = (y_1/f_1)/(y_2/f_2) \sim F_{f_1, f_2} \qquad 5.35$$

The F distribution is illustrated in Fig. 5.7 for various combinations of the two degrees of freedom that define it. The mean of a χ^2 distribution is close to the number of degrees of freedom so that if we divide by the number of degrees of freedom the mean will be close to 1. The expressions for the parameters of the F distribution are rather complicated: Table 5.9 gives the parameters for large values of f_1 and f_2 only.

5.5 CUMULATIVE DISTRIBUTION FUNCTIONS

For each of the distributions discussed in this chapter we can also determine the corresponding c.d.f. For discrete distributions we simply sum all the probabilities up to each observed value. For the families of eight children, given in Table 5.1, the estimated c.d.f. is

$$C_j = \sum_{i=0}^{j} f_i/F, \qquad 5.36$$

where F is the total number of families and f_i is the number of families having i boys. This is plotted in Fig. 5.8a. We can calculate the c.d.f. for the Poisson distribution of Prussian army officers (Table 5.4) in the same way and this is plotted in Fig. 5.8b.

The importance of the c.d.f. is that we can use it to determine the probability that an observable lies in a given range directly from graphs or tables. For example, from

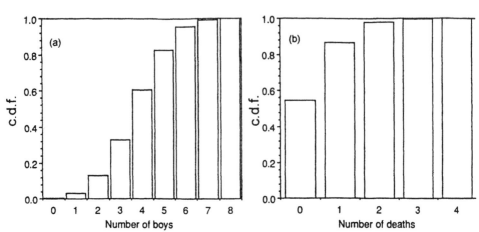

Figure 5.8 (a) The cumulative distribution function of the number of boys in families of eight children. (b) The cumulative distribution function of the number of Prussian army officers kicked to death by their horses.

Fig. 5.8 we see immediately that the probability that a family of eight children has four or fewer boys is 0.61. Similarly, the probability that none or one officer was kicked to death by their horses in any one year is 0.87.

For continuous distributions, the c.d.f. is the area under the p.d.f. up to each point (as we saw in section 3.2.3), so that for the normal distribution the value of the c.d.f. at x is the area under the p.d.f. from $-\infty$ to x. This is plotted for the standard normal distribution in Fig. 5.9. If we want to know the probability that a number falls in a

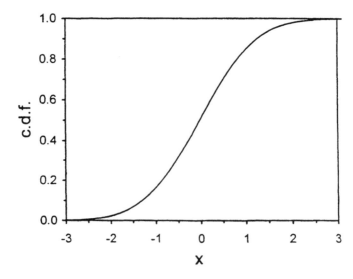

Figure 5.9 The cumulative distribution function for the standard normal distribution.

Table 5.10 Selected points from the c.d.f. of the standard normal distribution shown in Figure 5.9

x	$C(x)$	x	$C(x)$
− 3.29	0.0005	0.00	0.50
− 3.09	0.001	1.28	0.90
− 2.58	0.005	1.64	0.95
− 2.33	0.010	1.96	0.975
− 1.96	0.025	2.33	0.990
− 1.64	0.05	2.58	0.995
− 1.28	0.10	3.09	0.999
0.00	0.50	3.29	0.9995

certain range, say x_1 to x_2, then for both discrete and continuous distributions

$$P(x_1 < x < x_2) = \text{Area}(-\infty \text{ to } x_2) - \text{Area}(-\infty \text{ to } x_1) = C(x_2) - C(x_1), \quad 5.37$$

so that we can easily use the c.d.f. to find the probability that an observable falls into any given range.

Suppose we select a number from a normal distribution. The c.d.f. can be used to find the probability that the number we pick is less than 0, greater than 2, between 1 and 3 or falls into any range we choose. For example, Table 5.10 gives points from the c.d.f. for a standard normal distribution. If a number is drawn from a **standard** normal distribution, the probability that it is less than 0 is 0.5. If a number is drawn from **any** normal distribution the probability that it is less than the mean is 0.5. Similarly, the probability that a number drawn from a **standard** normal distribution is less than 1.96 is 0.975 (Table 5.10), so that the probability that a number chosen from **any** normal distribution is less than $\mu + 1.96\sigma$ is 0.975, or 97.5%, and so on. It also follows that the probability that a number drawn from a normal distribution is greater than $\mu + 1.96\sigma$ is 0.025, or 2.5%. Finally, we see that the probability that a number drawn from a normal distribution is greater than $\mu + 1.96\sigma$ or less than $\mu - 1.96\sigma$ is 0.05, or 5%.

Tables of c.d.f.s allow us to determine the probability that a number chosen from the appropriate distribution lies in any given range and we will use c.d.f.s extensively in this way. And, unlike the p.d.f., the c.d.f. is always dimensionless.

5.6 TABLES OF DISTRIBUTION FUNCTIONS

At the end of this book you will find tables of the c.d.f.s of the t, χ^2 and F distributions and you need to be able to use them. The row labelled P in the tables gives the value of the c.d.f., the first column gives the number of degrees of freedom and the values in the tables, which we call critical values, are the values of the statistics that we calculate. For example, the number in Table 10.1 corresponding to $P = 0.99$ with 6 degrees of freedom is 3.14. This tells us that if we measure a variable that follows the t distribution with 6 degrees of freedom, there is a 99% probability that it will be less than 3.14. Similarly, Table 10.2 tells us that if we measure a variable that follows the χ^2 distribution with 11 degrees of freedom, there is a 95% probability that it will be less than 19.68.

Tables of the F distribution are more extensive than the others because the F distribution has 2 separate degrees of freedom. Table 10.3, for example, gives the critical values for $P = 0.95$. For example, if we measure a variable that follows an F distribution with 6 and 12 degrees of freedom, there is a 95% probability that it will be less than 3.00. In the same way, Table 10.4 shows that if we measure a variable that follows an F distribution with 6 and 12 degrees of freedom there is a 99% probability that it will be less than 4.82.

You will notice that we do not give tables for the normal distribution. This is because the t distribution with an infinite number of degrees of freedom is identical to the normal distribution. Referring to Table 10.1, we see that if we choose a number from a normal distribution, there is a 99% probability that it will be less than 2.33.

5.7 SUMMARY

The binomial, Poisson, normal, χ^2, t, and F distributions provide the basis for most of what you need to know and we can summarize the conditions under which they hold as follows.

- Binomial: two outcomes, success or failure, probability of success is constant from one trial to the next and successive trials are independent.
- Poisson: limiting case of the binomial distribution when the probability of success in any one trial is very small but the number of trials is very large so that the mean number of successes remains finite.
- Normal: sum of a large number of independent variates.
- χ^2: sum of squares of normal variates.
- t: ratio of a normal variate to the square root of a χ^2 variate.
- F: ratio of two χ^2 variates.

There are other distributions of importance in biology. In particular, you may encounter the negative binomial distribution, which is often used to describe clumped distributions (Jeffers, 1978). For example, if a few cows in a herd have many ticks while most cows have few ticks, the distribution of the ticks on the cows can probably be described by a negative binomial distribution and one of its parameters can be used as a measure of the degree of clumping. The Poisson distribution turns out to be a particular case of the negative binomial distribution.

For the distributions we have discussed in detail and others that you will meet, it is important to be clear about the conditions under which they hold so that you can relate them directly to the biology of the problem that you are considering. In the next chapter we will use these distributions to see how we can make more rigorous and precise statements about differences between the results of our measurements and the predictions of our theories.

5.8 EXERCISES

1. Expand $(p + q)^2$ and $(p + q)^3$ and confirm that the successive terms give the probabilities listed in Table 5.1. Use Equation 5.2 to obtain the same result.

2. Use Equation 5.2 to verify the values for the expected numbers of families having four boys in Table 5.2.

3. Use Equation 5.30 to verify the values for the expected numbers of families having four boys in Table 5.6, assuming that the data are normally distributed with a mean of 4.118 and a standard deviation of 1.438.

4. Table 5.11, taken from Greenwood and Yule (1920), is an accident record for 647 women working in a munitions plant and gives the number of women having no accidents, one accident and so on. Calculate the mean number of accidents per woman. If the probability that any woman has an accident is small and is the same for all women, the numbers should follow a Poisson distribution. Calculate the expected frequencies and compare them with the observed frequencies.

Greenwood and Yule conclude that the number of women having three or more accidents is greater than expected and that there must have been two subgroups, one more accident prone than the other. Do you agree with their conclusions?

5. Larvae of the Azuki bean weevil (*Callosobruchus chinensis*) enter into beans (*Phaseolus radiatus*), feed and pupate inside them and then emerge through a hole (Utida, 1943). The number of holes per bean is therefore a good measure of the number of adults that have emerged. If the probability that any one bean is parasitized is small and does not depend on whether a bean has already been parasitized, the frequency distribution of the number of beans having 0, 1, 2, ... holes should follow a Poisson distribution. Use the data in Table 5.12 to calculate the expected numbers of beans having a given number of holes and compare these with the observed frequencies. The authors conclude that there are fewer beans with two or three holes than one would expect because weevils can identify and avoid beans that are already parasitized. Do you agree with this conclusion?

Table 5.11 The number of women having 1, 2, 3, 4 and 5 accidents in a five-week period

Number of accidents	Number of women
0	447
1	132
2	42
3	21
4	3
5	2

Table 5.12 The number of beans
having 1, 2 and 3 holes

Number of holes	Number of beans
0	61
1	50
2	1
3	0

6. Student (1907) counted the number of yeast cells in each of 400 squares of a haemocytometer with the results given in Table 5.13. Assuming that the number of cells follows a Poisson distribution, calculate the expected number of squares having any given number of cells in them and compare the observed and expected values. (This problem is analysed by Sokal and Rohlf, 1981, p. 84.)

7. Use the c.d.f. for the standard normal distribution (Fig. 5.9) to find the probability that a number chosen from a normal distribution with mean 1 and standard deviation 2 is (a) greater than 3; (b) less than 2; (c) lies between 2 and 3.

8. Use the tables at the end of this book to determine the probability that

   (i) a number chosen from a t distribution with 1 degree of freedom is (a) greater than 6.31; (b) less than -6.31; (c) lies between -6.31 and $+6.31$;
   (ii) a number chosen from a χ^2 distribution with 6 degrees of freedom is (a) less than 1.64; (b) greater than 12.59;
   (iii) a number chosen from an F distribution (a) with 3 and 4 degrees of freedom is greater than 6.59; (b) with 7 and 10 degrees of freedom is less than 6.06.

Table 5.13 The number of squares
having 0, 1, ... cells in them

Number of cells/square	Number of squares
0	75
1	103
2	121
3	54
4	30
5	13
6	2
7	1
8	0
9	1

6

Testing hypotheses

Science is the attempt to make the chaotic diversity of our sense experience correspond to a logically uniform system of thought. In this system single experiences must be correlated with the theoretic structure in such a way that the resulting coordination is unique and convincing.

The sense experiences are the given subject-matter. But the theory that shall interpret them is man-made. It is the result of an extremely laborious process of adaptation: hypothetical, never completely final, always subject to question and doubt.

The scientific way of forming concepts differs from that which we use in our daily life, not basically, but merely in the more precise definition of concepts and conclusions; more painstaking and systematic choice of experimental material; and greater logical economy. By this last we mean the effort to reduce all concepts and correlations to as few as possible logically independent basic concepts and axioms.

A. Einstein (1950)

The relationship between theory and experiment lies at the heart of modern science and has been the subject of intense debates in the history and philosophy of science. The Greeks and the Hindus established many of the most important branches of mathematics but did not develop a quantitative experimental science. The Romans were excellent engineers and the Chinese made the most extraordinary discoveries concerning the natural world but neither of them developed mathematical laws to describe natural phenomena. Kepler made a vital contribution to the development of modern science by showing that the orbits of the planets obeyed precise algebraic laws and Galileo's great contribution to science lay in the application of mathematics to the analysis of carefully conducted experiments. Indeed, Needham (1988) has said that 'Modern [as opposed to ancient or mediaeval] science is the mathematization of hypotheses about nature ··· combined with rigorous experimentation.'

Newton proposed the first scientific theory based solely on the formulation of a mathematical law, and although his law made it possible to predict the motion of the planets with great precision, many of his contemporaries rejected Newton's theory because it did not provide a mechanistic model of planetary motion. Darwin acknowledged that his theory of natural selection did not provide a mechanistic explanation of evolution, but he appealed to Newton's theory of gravitation in his defence, saying (Darwin, 1906, p. 657), 'It is no valid objection that science as yet throws no light on the far higher problem of the essence or origin of life. Who can explain what is the essence of the attraction of gravity? No one now objects to following out the results consequent on this unknown element of attraction, notwithstanding the fact that Leibniz formerly accused Newton of introducing occult qualities and miracles into philosophy'. So the debate continues. The 'idealists' claim primacy for theory, the 'materialists' claim primacy for experiment. Against the idealist position, we can argue that if you walk into a wall its reality is self-evident; against the materialist position, we can argue that the reason why the wall, viewed in a mirror, seems to be in front of us and not behind us is that our minds have already adopted the theoretical position that light travels in straight lines, and this is why we are deceived when we look in a mirror. Einstein (1978) said that 'The object of all science, whether natural science or psychology, is to coordinate our experiences into a logical system.'

It is important to remember that on the one hand we have our experiments and observations, on the other hand we have our theories and our deductions. The relationship between them is of a dialectical nature so that each informs the other. When we perform an experiment we do it to increase our understanding; when we build a theory we hope that it will reflect and clarify the results of our experiments. We do not perform our experiments in a 'theoretical vacuum'; there is always a theory or hypothesis we are trying to test. When we say that we should approach science with an open mind, we do not mean that we should have a blank mind, only that if our assumptions and hypotheses do not appear to be supported by the evidence, we should be willing and able to change our assumptions, invent new ones and maintain a flexible attitude. Indeed, whenever we look in a mirror, we should be reminded that even when we think that we are making no hypotheses, our subconscious is doing it for us. In this chapter we will consider the **testing of hypotheses** and **statistical inference** and see how these relate to statistics.

6.1 SIGNIFICANCE LEVELS AND THE POWER OF A TEST

Suppose we throw a coin 100 times and it lands heads on 55 occasions and tails on 45 occasions. How shall we decide if the coin is biased?

Let us begin by stating the problem carefully. We have an idea, I, that we want to test

$$I: \text{the coin is biased.} \qquad 6.1$$

We carry out an experiment that we hope will provide evidence for or against I and this gives us data, D

$$D: \text{55 heads and 45 tails.} \qquad 6.2$$

In science, it is all too easy to read into data what we want to see. We therefore proceed cautiously and make the **null** hypothesis

$$H: \text{the coin is } \textbf{not} \text{ biased.} \qquad 6.3$$

We now use our knowledge of statistics to decide if D seem reasonable if H is true. If D seem reasonable given H, we accept H and reject I (perhaps reluctantly). If D

seem unreasonable given H, we reject H, provisionally accept I and look for an explanation. To explain I we might, for example, examine the coin to see if it is bent. What
we need is a way to link the hypothesis, H, to the data, D. Making this link is what
biostatistics is all about. (We are discussing a simple problem in a formal way: when
we consider more complicated problems, proceeding formally should help.)

We have already worked out how to calculate the probability of getting 55 heads
and 45 tails in 100 throws of a coin using Equation 5.2 and the answer is

$$P(55 \mid 100) = \frac{n!}{k!(n-k)!} p^k q^{n-k}$$

$$= (100!/45!55!)0.5^{55}0.5^{45}$$

$$= 6.145 \times 10^{28} \times 7.889 \times 10^{-31} = 0.048. \qquad 6.4$$

so that this particular outcome is expected to occur only five times in every
100 experiments and all we can say is that it is rather improbable. However, even
the most probable outcome, which is 50 heads and 50 tails, only occurs eight times
in every 100 sets of 100 throws. Calculating the probability of getting the observed
result doesn't really help. If we think about this, it is obvious that if there are many
possible outcomes, even the most probable outcome will occur only very rarely;
looking at the probability that particular, individual events will occur is thus of little
value.

Let us try another approach. Even if our hypothesis is true and the coin is unbiased,
we would not expect to get precisely 50 heads and 50 tails, but rather to have results
which are close to 50 heads and 50 tails. We would surely agree that 50 heads and
50 tails is evidence in favour of the coin being unbiased while 100 heads and 0 tails
is evidence in favour of the coin being biased. We might then say that if we get
anywhere between 40 and 60 heads, we will regard the coin as unbiased, while if
we get less than 40 or more than 60 heads, we will regard it as biased. We could
calculate the probability of getting between 40 and 60 heads by adding the probabilities of each outcome in this range as we did in Equation 6.4. That rapidly becomes
tedious. However, we know that the expected number of heads is $pn = 0.5 \times 100 = 50$
and that the variance is $pqn = 25$ so that the standard deviation is 5. Since pqn is
sufficiently large, we can approximate the binomial distribution by a normal
distribution. With an expected mean of 50 and a standard deviation of 5, the probability
of getting between 40 and 60 heads is equal to the probability that a number chosen
from a normal distribution is within ± 2 standard deviations of the mean. Table 10.1
shows us that this probability is close to 95%. Since the probability of getting any
result is 1, the probability of getting less than 40 or more than 60 heads is $1 - 0.95 =$
0.05. In other words, if we repeated this experiment many times, we would expect
to get between 40 and 60 heads in 95% of the trials and less than 40 or more than
60 heads in 5% of the trials.

This test looks reasonable and this is how we will proceed, so let us tie it down
carefully. What we have done is to define an **acceptance range** (40 to 60 heads) and
a **rejection range** (0 to 39 and 61 to 100 heads). If the data, D, fall within the

acceptance range, we accept Hypothesis 6.3; if they fall in the rejection range, we reject it. Nevertheless, it is as well to bear in mind Gould's (1986) dictum that 'In science, "fact" can only mean "confirmed to such a degree that it would be perverse to withhold provisional assent" ', for even if the null hypothesis is true and the coin is unbiased, we know that the outcome may still be in the rejection range, and we would then reject the hypothesis even though it is true. We call this a **Type I error**. With the acceptance range of 40 to 60 heads, the probability of a Type I error is approximately 0.05 or 1 in 20. Since errors are undesirable, why don't we simply increase the acceptance range and in this way reduce the probability of making a Type I error? The trouble is that if we do this, the test becomes less sensitive. Suppose that we go to the extreme and include all results between 0 and 100 in our acceptance range. We will then never make a Type I error but we also will never reject Hypothesis 6.3 and we will conclude that even double-headed coins are unbiased! The danger now is that we accept our null hypothesis even when it is false and this is called a **Type II error**. We therefore have the following two kinds of error.

Type I error: reject when true.

Type II error: accept when false.   6.5

What we need is a balance: we want to keep the rejection range small to minimize Type I errors and keep the acceptance range small to minimize Type II errors. Because we cannot do both at the same time, our choice of range depends on our own judgement and the reasons for performing the experiment. For example, in English Law our hypothesis is that the defendant is innocent. The prosecution attempts to prove that the hypothesis of innocence is false. If an innocent person is punished, we are making a Type I error; if a guilty person is allowed to go free, we are making a Type II error. Since we prefer to make a Type II error and let a guilty person go free than to make a Type I error and make an innocent person suffer, the court insists that the case against the defendant be proved 'beyond reasonable doubt', which means that the rejection range is made very small and the probability of making a Type I error is minimized.

We now need more terminology and we call the probability of making a Type I error the **significance level** of the test. When we say that 'our data lead us to reject the hypothesis H at the 5% significance level', we are saying that although we believe that the hypothesis is indeed false, we acknowledge that we will be making a Type I error five times in every 100 experiments and that on these occasions the hypothesis really is true.

One minus the probability of making a Type II error we call the power of the test. Unfortunately, there is a major difficulty in specifying the power of a test because if we want to know the probability of accepting the null hypothesis when it is false we need to know what the alternative hypothesis is. The power of a test of one hypothesis can be defined only in relation to a second hypothesis. Let us call our hypothesis that the coin is unbiased the null hypothesis, H_0 and let us consider a second hypothesis, H_1, that the coin is biased in such a way that the probability that it falls heads is 0.6. We then spin our coin 100 times and record the number of

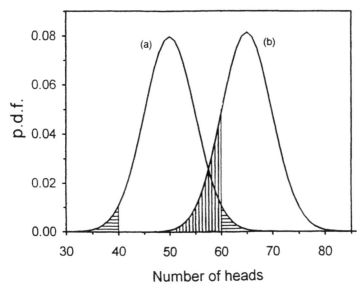

Figure 6.1 The two curves give the p.d.f.s for two hypotheses, (a) H_0: $P(H) = 0.5$ and (b) H_1: $P(H) = 0.65$. The acceptance range at a 5% significance level is from 40 to 60 heads. The horizontally shaded area is 0.05 and gives the probability of making a Type I error. The vertically shaded area is 0.16 and gives the probability of making a Type II error. The power of the test is $1 - 0.16 = 0.84$.

times it shows heads. Figure 6.1 shows the probabilities of the various possible outcomes under each of the two hypotheses. Suppose we now set the acceptance range to include any outcome in which the number of heads is between 40 and 60, and set the rejection range to include any outcome in which the number of heads is less than 40 or greater than 60. The horizontally shaded area in Fig. 6.1 is the probability of making a Type I error, in this case 0.05, and determines the significance level. The vertically shaded area in Fig. 6.1 is the probability of making a Type II error, in this case 0.16, and accepting H_0 when H_1 is true and H_0 is false. Since we want the power of the test to be high when the probability of making Type II error is small, we define the power of the test as $1 - 0.16 = 0.84$. If we reduce the significance level to increase protection against Type I errors by widening the acceptance range, we also reduce the power of the test, giving less protection against Type II errors.

Now if we are in that rather rare situation in which we have only two possible hypotheses, then we can talk about both the significance level and the power of the test and choose our acceptance and rejection ranges accordingly. More often we will have only our null hypothesis to test. If we decide to reject it, we may think of an alternative explanation, formulate a new hypothesis and then test the new one, but we will generally be concerned with one hypothesis at a time. For this reason, we will nearly always specify the significance level of our tests rather than their power. However, when we plan experiments, we can use the power of a test to decide, in advance, how big our samples should be. For example, we have seen in the trial of the polio vaccine that with 200 000 children in each group there was little doubt about the effectiveness of the vaccine, but we have also seen that with only 20 000 children in each group, the effect of the vaccine may have been lost in the statistical noise. In section 7.8.2 we will show how we can use the power of a test to help us to decide how many children we need to include in such a study.

6.1.1 Summary

Before we perform an experiment we must have an idea that we are trying to test. The danger is that we deceive ourselves and read into the data what we want to see there rather than what is there. We therefore err on the side of caution by making the null hypothesis that the effect is not there and then we try to disprove the null hypothesis. If we succeed we look more favourably on the idea we are testing, if we fail we look less favourably on the idea we are testing.

Formally, we proceed as follows:

- We have an idea we want to examine (the coin is biased).
- We make a series of measurements (throw the coin 100 times) that give us data, D, (H, H, H, T, H, T, T, H, T ...).
- We make an hypothesis, H (the coin is not biased).
- We calculate a test statistic, x (65 heads, say), based on D.
- Assuming that H is true, we use our knowledge of statistics to decide on the distribution that x should follow (binomial with $p = 0.5$, $n = 100$).
- We decide on a significance level P(5%, say) and determine critical values (40 and 60 heads) that define the acceptance range (40–60 heads), and the rejection range (0–39 and 61–100 heads) at this level of significance.
- If x (65) falls in the acceptance range (40–60) then **we provisionally accept H**, and conclude that the coin is not biased, remembering that more extensive data (more throws of the coin) may lead us to change our minds in the future.
- If x (65) falls in the rejection range (0–39 and 61–100 heads) **we reject H at the P% significance level** and we conclude that the coin is probably biased, remembering that although we are confident about rejecting H there is still a probability P (5%) that we are making a Type I error and the coin is not in fact biased (since even an unbiased coin could produce 65 heads and 35 tails).

This may seem pedantic but science proceeds by the accumulation of very many small steps and it is vital that we do not lose our way or allow ourselves to be seduced into seeing what we want to see.

6.2 THE χ^2 TEST

In Chapter 5 we examined the distribution of the number of boys in families of eight children and compared the observed distribution with the distribution we expected if the assumptions of the binomial distribution were satisfied. In particular we wanted to know if there was any evidence that large families tend to produce more boys or girls than expected, for if this is so then some parents may have a tendency to produce children of one sex. We also looked at the number of plants in Mendel's experiments having various pairs of characteristics and we wanted to know if the observed numbers supported Mendel's theory.

When we compare observed and expected numbers, a convenient and powerful test is the χ^2 test based on the χ^2 distribution. Before we develop the test, remember that we will start by assuming that 'nothing is going on': our coins are unbiased, there is no tendency to produce boys rather than girls, there is no tendency to produce children of the same sex, the polio vaccine does not prevent polio, and so on. Our initial hypothesis H is therefore referred to as the **null hypothesis**.

For our study of the number of boys in families of eight children, our null hypothesis is that there is no propensity to produce children of the same sex so that the observed numbers of boys and girls should agree with the predictions of the binomial distribution calculated in Table 5.1 and repeated in Table 6.1. We can then restate H as follows:

H: the observed frequencies do not differ significantly

from those expected for a binomial distribution. 6.6

Let us first look at $O - E$, the difference between the observed frequencies O, and the expected frequencies E. If H_0 is true, the expected value of $O - E$ is 0. Now, if the probability of any given outcome, two boys and six girls, say, is small, O, the observed number of times each outcome occurs will follow a Poisson distribution so that the variance of O is equal to its expected value E. $(O - E)/\sqrt{E}$ will therefore have an expected value of 0 and a variance of 1. Provided E is not too small, $(O - E)/\sqrt{E}$ will be approximately normally distributed. Therefore, if we square $(O - E)/\sqrt{E}$ for each pair of observed and expected values and add them all together, this variable will come from a χ^2 distribution, provided our null hypothesis is true, so that

$$\sum_{i=1}^{n} (O_i - E_i)^2/E_i \sim \chi_m^2 \qquad 6.7$$

Table 6.1 Calculation of χ^2 from the observed and expected numbers of boys in families with eight children. The sum of the χ^2 values is 92

Number of boys	Observed frequency	Expected frequency	Difference	χ^2
0	215	165	+ 50	15.2
1	1 485	1 402	+ 83	4.9
2	5 331	5 203	+ 128	3.2
3	10 649	11 035	− 386	13.5
4	14 959	14 628	+ 331	7.5
5	11 929	12 410	− 481	18.6
6	6 678	6 580	+ 98	1.5
7	2 092	1 994	+ 98	4.8
8	342	264	+ 78	23.0

where the sum is over all our observations and m is the number of degrees of freedom. (This result is more subtle than it appears and is discussed further in the Appendix, section 6.5.1. In fact we only need to assume that O follows a binomial distribution rather than the more restrictive assumption that O follows a Poisson distribution.) We can then test Hypothesis 6.6 by comparing the left-hand side of Equation 6.7 with a suitable acceptance range chosen from the χ^2 distribution with m degrees of freedom.

We have to be careful with the number of degrees of freedom. For our families with eight children, we start off with nine possible outcomes: 0, 1, 2...8. Once eight of the probabilities have been fixed, the last is also fixed since they must add up to 1, so this removes 1 degree of freedom. However, we have also used the data to estimate the probability that any one child is a boy, so we lose another degree of freedom. The total number of degrees of freedom is therefore 7 and we use the χ^2 distribution with 7 degrees of freedom to determine the critical value that defines the acceptance range.

Table 6.1 shows the data of Table 5.2 with the contributions to the χ^2 distribution calculated for each observation. Adding all the numbers in the last column of Table 6.1 gives a value of 92. The next thing we need to do is to decide on our acceptance range. Suppose we are willing to risk making a Type I error by rejecting the hypothesis even though it is true, about 1 time in 20. We then look up the 5% significance level for a χ^2 distribution with 7 degrees of freedom and Table 10.2 tells us that the critical value is 14.1, so our acceptance range is from 0 to 14.1 and our rejection range is anything greater than 14.1. The χ^2 statistic is 92, well outside the acceptance range, and we can decisively reject our hypothesis that the observed frequencies in Table 6.1 follow a binomial distribution.

To summarize, we want to know if there is a propensity for some families to produce children of the same sex so we start with the null hypothesis that there is no propensity for some families to produce runs of boys and others to produce runs of girls. We then model the results we expect to find if our null hypothesis is consistent with the data. For the example of boys and girls in Saxony, we find reasonably good agreement between the observed and expected numbers, but because the sample size is very large, the data are very precise and our statistical test is able to pick up very small differences. Although the observed distribution is close to the distribution predicted using our null hypothesis, we find that there are small but significant differences. The data appear to suggest that there is a slight genetic propensity to produce children of the same sex. To pursue the matter further we would examine the deviations from the predictions, formulate new hypotheses, make new models and hope eventually to find a model that not only fits the data but also makes biological sense.

6.2.1 Mendel's peas

Mendel found that the observed ratios of dominant to recessive characters in the F_2 generations of his peas were always close to 3 to 1 (Table 2.2). We can now use the χ^2 test to see whether the deviations from the expected ratio of 3 to 1 are small enough to be attributed to chance alone. Our hypothesis in each case is then

H: The ratio of the number of dominants to recessives

in the F_2 generation is $3:1$. 6.8

For each experiment in Table 2.2 we take the total number of peas (7324 for the comparison of round and wrinkled seeds) and calculate the expected number of dominants ($7324 \times 0.75 = 5493$ for round seeds) and recessives ($7324 \times 0.25 = 1831$ for wrinkled seeds) on the hypothesis that the expected ratio of dominants to recessives is $3:1$. We then use Equation 6.7 to calculate the values of χ^2, so that for the experiment involving round and wrinkled seeds we have

$$\chi^2 = \Sigma (O - E)^2 / E$$
$$= (5474 - 5493)^2 / 5493$$
$$+ (1850 - 1831)^2 / 1831$$
$$= 0.263. \qquad 6.9$$

The χ^2 statistics for each of Mendel's experiments are given in Table 6.2. In each experiment we have two measurements, the number of plants showing the dominant character and the number showing the recessive character, but we have had to use the total number of plants to calculate the expected frequencies, so we lose 1 degree of freedom, and we need to compare each statistic with a χ^2 distribution with 1 degree of freedom.

To see whether each of these numbers could reasonably come from a χ^2 distribution with 1 degree of freedom, we let our acceptance range correspond to the lower 95% of the distribution, so that our significance level is 5%. From Table 10.2, the critical value of χ^2_1 at the 5% significance level is 3.84, so that the values of χ^2 in Table 6.2 are all well within the acceptance range. In fact, they are all so well within the acceptance range that they look suspiciously good and we may begin to wonder if Mendel did not in fact 'massage' his data to improve the agreement between the

Table 6.2 Values of χ^2 for Mendel's experiments on peas calculated from the data given in Table 2.2

Characters	χ^2
Round vs wrinkled seeds	0.263
Yellow vs green seeds	0.015
Purple vs white flowers	0.391
Smooth vs constricted pods	0.064
Axial vs terminal flowers	0.350
Green vs yellow unripe pods	0.451
Tall vs dwarf stems	0.607

experimental values and his expectations. Let us, therefore, make a new hypothesis,

H: The ratio of dominants to recessive is 3:1 and

Mendel was an honest scientist,   6.10

and see if his data gives us cause to reject the hypothesis. Our suspicion has been aroused because of the consistently small values of χ^2 obtained in Table 6.2 and we want to know if seven such small values could reasonably have arisen by chance.

Usually our acceptance range would include all values less than the critical value, because significant differences between observed and expected values will give large values of χ^2 that we treat with suspicion. However, if Mendel tried to improve the agreement between his observed and expected values, it is small values of χ^2 that we treat with suspicion and we now choose our acceptance range so that the observed numbers should be **greater** than the critical value with 95% probability and **less** than the critical value with 5% probability. In other words, we take the bottom 5% rather than the top 5% of the distribution as our rejection range. For a χ^2 distribution with 1 degree of freedom, the bottom 5% of the distribution corresponds to values between 0 and 0.004 (Table 10.2) and all of the values in Table 6.2 lie within the acceptance range so that we cannot reject the hypothesis that Mendel was an honest scientist at the 5% significance level.

The fact that we have seven values of χ^2, which taken together seem to be consistently small, leads us to wonder if we can combine them and obtain a more sensitive test of our hypothesis concerning Mendel's honesty. Adding the χ^2 values (each of which has 1 degree of freedom) gives 2.14 and if Hypothesis 6.10 holds, this will be a random sample from a χ^2 distribution with 7 degrees of freedom. For a χ^2 distribution with 7 degrees of freedom, the critical value for the bottom 5% of the distribution is 2.17 (Table 10.2) and since this is slightly greater than 2.14, we **can** reject the hypothesis that Mendel was an honest scientist at the 5% significance level. The probability that we are making a Type I error (rejecting the hypothesis when true) and accusing Mendel falsely is only 1 in 20.

The debate as to what Mendel actually did is complex but the evidence against Mendel can be made stronger than it has been made here. Fisher carried out a detailed analysis of all of Mendel's published data and concluded that the data are too good to be true (Fisher, 1936). However, saying that Mendel 'cheated' does not get us very far because his theory was correct: he had, in fact, made one of the most important scientific discoveries of all time. If his work had been understood and appreciated, the study of evolution and inheritance might have been advanced by 30 years. What we really want to know is how and why did he manipulate his data?

On the evidence available to him Fisher concluded that 'after examining various possibilites, I have no doubt that Mendel was deceived by a gardening assistant who knew too well what his principle expected from each trial' (Fisher, 1965). Blaming an unnamed assistant would seem to be wishful thinking on Fisher's part and an attempt to protect Mendel's honour. It seems to me equally likely that Mendel deceived himself: every scientist tends to repeat experiments that give the 'wrong' answer, stopping as soon as the 'correct' answer turns up.

Even if Mendel's data are falsely presented, various interpretations are possible. It appears that Mendel did not realize that he had discovered the genetic basis of inheritance and that his work on peas could be generalized to include the entire living world. He seems rather to have believed that the numerical proportions he discovered for the inheritance of characteristics in peas would have to be determined separately for each species of plant or animal. Mendel appears not to have speculated on the mechanism that might have given rise to his observations: if he had, he would surely have been led to postulate pairs of units (genes), one inherited from the male, the other inherited from the female. In other words, Mendel may have been concerned in his paper primarily to demonstrate his methodology and to educate others, and may have thought that the precise numbers were unimportant. If so, it may have seemed entirely reasonable to select those numbers that best demonstrated the result he was trying to establish. (Broad and Wade (1985) have written about fraud in science and Gould (1981) has addressed such questions in relation to intelligence testing.) I have done something similar in this book by selecting examples that match my expectations while avoiding those that are overly complicated. As Gould (1980) says, 'Science is not an objective, truth directed machine, but a quintessentially human activity, affected by passions, hopes and cultural biases.'

6.3 CONTINGENCY TABLES

Once we agree that Mendel's theory of inheritance is essentially correct and that the various characteristics are determined by what we now call genes, we are led to wonder whether the genes for different characters are related: perhaps the genes that confer 'roundness' and 'yellowness' on the seeds are linked. The question as to whether two properties are related arises frequently in biology and is amenable to analysis using a χ^2 test.

Mendel reported the results of crosses involving more than one pair of characters, but since he had not developed a mechanistic explanation of inheritance, he did not imagine that different pairs of characters might be linked—he assumed that the various characters were independent and then proceeded to show that the various combinations occurred in the appropriate proportions. It is now known that of the genes studied by Mendel only two are linked: those that determine the colour of the unripe pods and those that determine the length of the stem (Lambrecht, 1961). With our present-day knowledge of genes and chromosomes, let us test Mendel's data for linkage. To do this we use **contingency tables**, such as Table 2.5 repeated here as Table 6.3. It is called a 'contingency table' because we can use it to calculate the probability that a seed is round contingent on it also being green, and vice versa: in other words, we can use it to test conditional probabilities.

In section 2.2.3 we estimated, for example, the probability that a seed will be round given that it is yellow, $\hat{P}(\text{round}|\text{yellow})$, and showed that this is almost the same as $\hat{P}(\text{round}|\text{green})$. We then argued that since the probability of it being round is the same whether it is yellow or green, the gene for shape and the gene for colour must be independent. But we need to quantify this observation and decide if the small differences in the two probabilities are significant or not.

Table 6.3 Mendel's data on the distribution of seed colour and shape

	Yellow	Green	Total
Round	315	108	**423**
Wrinkled	101	32	**133**
Total	**416**	**140**	**556**

To develop our χ^2 test for contingency tables, we again start from the hypothesis we want to test, in this case

H: The traits for seed colour and shape are independent.   6.11

Given this assumption, we can use our statistical theory to calculate the expected frequencies for seeds of each colour and shape and then use a χ^2 test to compare the observed and expected frequencies.

For Mendel's data, we want to calculate the expected frequency in each cell of Table 6.3 under Hypothesis 6.11, taking the row and column totals as given. As we saw in section 2.2.3,

$$\hat{P}(\text{round}) = 423/556 \quad \hat{P}(\text{yellow}) = 416/556. \qquad 6.12$$

and we know from Equation 2.17 that if 'roundness' and 'yellowness' are statistically independent, then

$$\hat{P}(\text{round and yellow}) = \hat{P}(\text{round}) \times \hat{P}(\text{yellow}) = 423 \times 416/556^2 = 0.569. \quad 6.13$$

Since we have a total of 556 plants, the expected number of plants whose seeds are both round and yellow is $0.569 \times 556 = 316.5$, while the observed number is 315. The contribution to the χ^2 sum is $(315 - 316.5)^2/316.5 = 0.0070$. Repeating this for the other three combinations of colour and shape gives the expected frequencies and contributions to the χ^2 sum shown in Table 6.4. Adding together the four contributions to the χ^2 sum gives 0.116.

Once again we need to decide how many degrees of freedom to use in our χ^2 test. We started with four measurements but we used the total number of seeds (556), the marginal number of round seeds (423), and the marginal number of yellow seeds (416), so that we have lost 3 degrees of freedom. We do not count the marginal number of wrinkled seeds or the marginal number of green seeds, since once we have fixed the three numbers mentioned above, these two are also fixed and we lose nothing further by using them. We are therefore left with $4 - 3 = 1$ degree of freedom. (If you calculate $O - E$ for each cell in Tables 6.3 and 6.4, you will get ± 1.49 for each of them, showing that although we have four differences, there is really only one independent difference.)

If we choose to test the experiment at the 5% significance level, the acceptance range for χ^2_1 is from 0 to 3.84 (Table 10.2), and we see that Mendel's result, 0.116, falls well within this range (perhaps too well within it), and we accept the hypothesis that the genes for shape and colour are independent.

Table 6.4 The expected frequencies and the contributions to the χ^2 sum for the different combinations of seed colour and shape corresponding to the observed values in Table 6.3. The χ^2 values add up to 0.116

| | Expected | | χ^2 | |
	Yellow	Green	Yellow	Green
Round	316.49	106.51	0.007	0.021
Wrinkled	99.51	33.49	0.022	0.066

Now we have ignored some of the information inherent in Mendel's data because his theory predicts that the marginal totals should be in the ratio of 3 to 1, and instead of using the observed marginal totals to calculate the expected cell frequencies, we could have used his theory. If we make use of his theory, our hypothesis is slightly different and we have

H: The ratio of dominants to recessives is 3 to 1 and the

genes for roundness and colour are independent.   6.14

The probability that a seed is both round and yellow under Hypothesis 6.14 is then $0.75 \times 0.75 = 0.5625$, and the expected cell frequency for plants whose seeds are both round and yellow is $0.5625 \times 556 = 312.8$, which we see is slightly different from the previous expected value of 316.5. We again calculate the expected frequencies for each cell in our table and use these to calculate the value of χ^2, which turns out to be 0.47, a larger number than we had before. This time, however, we have lost only 1 degree of freedom in using the total number of plants to determine the expected frequencies, so we are left with $4 - 1 = 3$ degrees of freedom. We must now compare our result with a χ^2 distribution with 3 degrees of freedom, for which the acceptance range at the 5% significance level is from 0 to 7.81 (Table 10.2). We see that we are again well within the acceptance range and we should accept the Hypothesis 6.14 that the ratio of dominants to recessives is 3 to 1 and that the genes for shape and colour are independent.

6.3.1 Multiway contingency tables

There is no reason to limit ourselves to only two categories for each variable in our contingency table. The extension to more than two is straightforward and the rules for a contingency table with *R* rows and *C* columns are as follows.

If we need to use the marginal totals to calculate the expected cell frequencies (as in our analysis based on Hypothesis 6.11), we proceed as follows.

1. Calculate the expected cell frequencies, by multiplying the appropriate row total by the appropriate column total and dividing by the grand total.
2. Carry out a χ^2 test with $(R - 1) \times (C - 1)$ degrees of freedom.

If we have a theory that predicts the marginal totals (as in our second analysis based on Hypothesis 6.14), we proceed as follows.

1. Calculate the expected marginal totals from the grand total using the theory, and then calculate the expected cell frequencies as in the previous case.
2. Carry out a χ^2 test with $RC - 1$ degrees of freedom.

Convince yourself of the validity of Rule 1. To understand Rule 2 in the first case, note that we start with RC frequencies. We lose 1 degree of freedom when we use the grand total. With $R - 1$ row totals and the grand total we can calculate the remaining row total. We therefore lose another $R - 1$ degrees of freedom when we use the row totals to calculate cell frequencies. Similarly, we lose $C - 1$ degrees of freedom when we use the column totals. This leaves $RC - (R - 1) - (C - 1) - 1 = (R - 1) \times (C - 1)$ degrees of freedom. For Rule 2 in the second case we only use the grand total and hence lose only 1 degree of freedom, leaving $RC - 1$ degrees of freedom.

6.3.2 Corrections for small numbers

A critical assumption in our development of the χ^2 test is that we can approximate a binomial distribution by a normal distribution. If the numbers are small, this can lead to significant bias in the results. Snedecor and Cochran (1989, p. 77) give the following rules for applying the χ^2 test to small samples.

- No class expectation should be less than 1.
- Two expected values may be close to 1 if most of the others exceed 5.
- If necessary, classes should be combined to meet these rules.

For a 2×2 contingency table, a conservative recommendation (Sokal and Rohlf, 1981, p. 711) is that no expected number should be less than 5 and a correction factor, known as Yate's correction (Bulmer, 1979, p. 163; Hays, 1988, p. 774), can be included in 2×2 tables by replacing Equation 6.7 with

$$\chi^2 = \Sigma (|O - E| - 0.5)^2 / E \qquad\qquad 6.15$$

where $|a|$ indicates the absolute value of a. (The absolute value is the number with a positive sign so that $|-3| = |+3| = +3$.) Alternatively, provided that the numbers in each cell are not too large we can calculate the probabilities of all possible outcomes exactly which is the basis of Fisher's exact test (Siegel and Castellan, 1988, p. 103; Hays, 1988, p. 781).

6.4 SUMMARY

When you apply statistics to the analysis of your data, there will always be some idea or theory that you wish to test and of course it is precisely in the formulation of such ideas or theories that the creativity of science lies. However, we choose to proceed in a rather indirect way. If we suspect that a coin is biased we assume initially that it is not and then try to disprove our null hypothesis. Given our null hypothesis, we use our knowledge of biology and statistics to decide if the observed data seem reasonable if the null hypothesis is indeed true. If the observed data seem reasonable, we accept our null hypothesis and conclude, provisionally, that our idea cannot be supported by the data. It is still open to us to design a more subtle experiment, carry out more trials or in some way try to tease out the information that we believe is there. If the observed data do not seem reasonable given the nulll hypothesis, we conclude, provisionally, that our idea is supported by the data. It may still be the case that the observed effects could be explained on the basis of some other idea, one that is quite unrelated to the one we are proposing. So we try to test our ideas in as many different ways as possible and if we still come up with the same theory we begin to believe it. At the end of the day, all that matters is that we can provide a sound and convincing biological explanation for the effects that we observe. The mathematics is the handmaiden to the biology that we hope to understand.

The χ^2 statistic, which we have examined at some length, is only one of many statistics that we can derive from our experimental data. In the next three chapters we will examine other situations and other statistics that we can use to test hypotheses about our data.

6.5 APPENDIX

6.5.1 The χ^2 test

Consider an experiment in which there are only two possible outcomes such as heads or tails, tall or short plants, boys or girls, and so on. Let N be the total number of events (throws of the coin, plants sampled, children) of which a fall into the first category (heads, tall plants, boys), while b fall into the second category (tails, short plants, girls). Let α be the probability that a occurs and let β be the probability that b occurs. Then

$$a + b = N \quad \alpha + \beta = 1 \qquad \qquad 6.16$$

$$E(a) = \alpha N \quad V(a) = \alpha \beta N \qquad \qquad 6.17$$

$$E(b) = \beta N \quad V(b) = \beta \alpha N. \qquad \qquad 6.18$$

Now

$$\chi^2 = \Sigma (O - E)^2 / E$$
$$= (a - \alpha N)^2 / \alpha N + (b - \beta N)^2 / \beta N$$
$$= \{\beta(a - \alpha N)^2 + \alpha(b - \beta N)^2\} / \alpha \beta N. \qquad \qquad 6.19$$

But

$$b - \beta N = (N - a) - (1 - \alpha)N = \alpha N - a, \qquad \qquad 6.20$$

so that substituting for $b - \beta N$ in equation 6.19,

$$\chi^2 = (a - \alpha N)^2 (\beta + \alpha) / \alpha \beta N = (a - \alpha N)^2 / \alpha \beta N. \qquad \qquad 6.21$$

Now $E(a - \alpha N) = 0$ and $V(a - \alpha N) = \alpha\beta N$ so that $(a - \alpha N)^2/\alpha\beta N$ has mean zero and variance 1. Provided αN is sufficiently large we can assume that its distribution is approximately normal, so that χ^2 given by Equation 6.21 belongs to a χ^2 distribution with 1 degree of freedom.

The key point to note is that although the variance pqN of a binomially distributed variable is **not** equal to the expected value pN, we nevertheless divide by the expected value in Equation 6.19. The sum then reduces to the correct expression for a χ^2 distribution with the number of degrees of freedom equal to the number of degrees of freedom in the data. In other words, Equation 6.7 holds for binomially distributed observations and p need not be small as would be the case if it only held for observations following a Poisson distribution as implied in the text.

It is still necessary however for each expected value to be sufficiently large for the assumption of normality to hold and experience shows that this does hold provided the conditions noted in section 6.3.2 are satisfied.

6.6 EXERCISES

1. Carry out χ^2 tests on the hypothetical data in Table 1.4 and decide which numbers confirm Mendel's theory and at what significance level. How good were your guesses?

2. The data of Table 1.2 for the results of the trial of the polio vaccine are repeated as a contingency table in Table 6.5. Carry out a χ^2 test to decide if the incidence of polio is related to use of the vaccine.

3. Carry out a χ^2 test on the data of Table 5.4 and decide whether to accept or reject the hypothesis that the number of deaths follows a Poisson distribution. (Note the recommendations in section 6.3.2.)

4. Carry out a χ^2 test of Bateson's data in Table 2.6 using both the marginal totals and Mendel's theory to predict the cell frequencies. Are we justified in concluding that in Bateson's experiment the shape of the pollen is related to the colour of the flowers?

5. In an experiment to investigate the possible protective effects of vitamin C against rabies 98 guinea-pigs were given a small dose of fixed rabies virus. Of these 48 were also given vitamin C, the rest were not. Table 6.6 gives the number that died or survived under each treatment (Banic, 1975). Deaths occurred in 35% of the treated and

Table 6.5 The number of children that developed polio and that remained healthy after receiving either the vaccine or a placebo

	Paralytic cases	Healthy cases
Vaccine	33	200 712
Placebo	115	201 114

Table 6.6 The number of treated and untreated guinea-pigs that died or survived in an experiment on the efficacy of vitamin C in treating rabies

Treatment	No. of deaths	No. of survivors
Vitamin C	17	31
Control	35	15

70% of the control animals. Carry out a χ^2 test with and without Yate's correction to decide if vitamin C is effective against rabies.

6. In August 1989 *Newsweek* published a report on a US government-funded study that showed that AZT (azidothymidine) helps to slow down the development of the diseases AIDS (acquired immune deficiency syndrome) (*Newsweek*, 1989). Either AZT or a placebo was given to 713 people. At the time of writing, 14 of those taking AZT had developed AIDS, while 36 of those given the placebo had developed AIDS. Do these data encourage us to believe that AZT slows the development of AIDS? (Hint: Without knowing how many received the drug and how many received the placebo, we can draw no conclusions. Let us therefore assume that about half were given the drug and half the placebo.)

7. The shell of the snail *Cepaea nemoralis* can be yellow, brown or pink and it can be banded or uniform. Cain and Sheppard (1952, 1954) showed that shell colour is adaptive in that it provides camouflage against attack by thrushes. In habitats providing a uniform background the shells tend to be uniform in colour; in habitats providing a varied background the shells tend to be banded. To confirm that this was the result of selective predation by thrushes, Sheppard (1951) counted the numbers of banded and unbanded snails that were living and the numbers that were killed by thrushes in a certain area. Since a thrush breaks the snail shell by smashing it on a rock, it was possible to sample the number killed by thrushes in each class. The results are given in Table 6.7. Use a χ^2 test to decide if there is a significant association between 'bandedness' and being killed by a thrush. Do these data support Cain and Sheppard's predation hypothesis?

Table 6.7 The numbers of banded and unbanded snails that were living and that were killed by thrushes

	Banded	Unbanded
Living	264	296
Killed	486	377

7

Comparisons

It's proper to begin with the regular facts, but after a rule has been established beyond all doubt, the facts in conformity with it become dull because they no longer teach us anything new. Then it's the exception that becomes important. We seek not resemblances but differences, choose the most accentuated differences because they're the most striking and also the most instructive.   R.M. Pirsig (1980)

When we analyse data we have an hypothesis that we wish to test. We then use our knowledge of statistics to give us a measure of the probability that the observed data will arise in an experiment if the hypothesis is true. So far we have considered situations in which we want to know only if the measured data could reasonably have arisen from a particular hypothetical distribution. In practice, life is usually more complicated than this. We may have a number of different variables, perhaps rainfall and temperature, each of which might affect the outcome of our experiment, which might be the growth of a crop, and we want to be able to test each of them. We may even find that the effect of one variable depends on the particular values of another so that high temperatures might increase crop growth when it is wet but lead to decrease in crop growth when it is dry.

In this chapter we will consider problems that arise when we compare measurements made under several different conditions. We will consider some aspects of the design of experiments and this will lead us into the chapters on the analysis of variance and regression where we formalize the way in which we design experiments and analyse data.

7.1 ONE- AND TWO-TAILED TESTS

In Chapter 6 we compared the observed and expected number of boys in families with eight children. To do this we used a χ^2 test and since a failure of the hypothesis would lead to large values of χ^2, we took as our acceptance range (at the 5% significance level) all values less than 11.1 and our rejection range corresponded to the top 5% of the distribution. To test Mendel's data on peas, we again argued that failure of the hypothesis would lead to large values of χ^2 and our rejection range was again the top 5% of the distribution. However, when we set out to test Mendel's honesty, we argued that if he had cheated, the observed results would have been

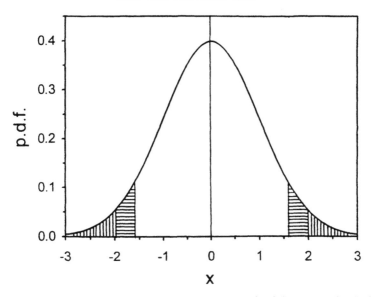

Figure 7.1 The p.d.f. for a standard normal distribution. Each of the separately shaded areas contains 2.5% of the total area.

smaller than expected, and so we took as our acceptance range (at the 5% significance level) all values greater than 2.26 and our rejection range then corresponded to the bottom 5% of the distribution. On still other occasions we will want to know if two observations differ in either direction and on these occasions we will want our rejection range to include both tails of the distribution.

To illustrate this, Fig. 7.1 shows the p.d.f. for a standard normal distribution. Each of the separately shaded areas contains 2.5% of the total area. Suppose we now measure two numbers, let us call them a and b, and take as our null hypothesis that their difference belongs to a standard normal distribution. Then we can set up various alternative hypotheses as indicated in Table 7.1. If the alternative hypothesis is that $a > b$, we take as the rejection range the two shaded areas on the right-hand side of Fig. 7.1; if it is that $a < b$, we take the two shaded areas on the left-hand side of Fig. 7.1; if it is that $a \neq b$, we take as the rejection range the two vertically shaded areas at the two ends of Fig. 7.1. The point to bear in mind is that in all three cases given in Table 7.1, the significance level is the same but the acceptance

Table 7.1 Tests of the hypothesis $a = b$ at the 5% significance level, assuming that $a - b$ follows a standard normal distribution, for various alternative hypotheses

Null Hypothesis	Alternative hypothesis	Rejection range for $a - b$
$a = b$	$a > b$	1.65 to ∞
$a = b$	$a < b$	$- \infty$ to 1.65
$a = b$	$a \neq b$	$- \infty$ to $- 1.96$ & 1.96 to ∞

range depends on the alternative hypothesis so that we are effectively choosing the acceptance range in such a way as to maximize the power of the test for a given alternative hypothesis.

7.2 COMPARISON OF MEANS

Suppose that we want to compare the results of two measurements. The first thing we will probably want to know is whether the means are significantly different. This is essentially what we did in section 5.2.2 in our study of the effects of the polio vaccine where we noted that the probability of a vaccinated child developing polio was 16.4 ± 2.9 per 100 000, while the probability of an unvaccinated child developing the disease was 57.1 ± 5.3 per 100 000. Using the result (Appendix, section 4.5.2) that for two independent random variables, the variance of their sum or difference is the sum of their variance, we were able to calculate the standard deviation of the difference as

$$s = (2.9^2 + 5.3^2)^{1/2} = 6.0. \qquad\qquad 7.1$$

and we noted that this was much less than 40.7, the reduction in the number of children per 100 000 contracting polio when they were given the vaccination.

Now we can be more precise in our analysis. The hypothesis we want to test is that the vaccine brings about a significant reduction in the number of children who contract polio and so we use a one-tailed test. Since there are a large number of trials and the numbers of children contracting polio are not too small, we can approximate the distribution of the difference in the numbers contracting polio under the two treatments by a normal distribution. For a **standard** normal distribution, 0.1% of the area under the curve is above 3.09 (Table 10.1), so that for a normal distribution with mean zero, the hypothesis we are testing, and standard deviation s, 0.1% of the area is above $3.09s = 3.09 \times 6.0 = 18.5$. The difference between the numbers of children contracting polio in the two groups, 40.7, is well outside the acceptance range at the 0.1% significance level so that we are able to conclude that the vaccine does indeed bring about a significant reduction in the incidence of polio.

In this example we have used a one-tailed test. Suppose, however, we had two different vaccines and we wanted to know if there was any significant difference between them. We would not know in advance which was the better of the two and we would then use a two-tailed test. Our acceptance range at a 0.1% significance level would then be $-3.29s$ to $+3.29s$ (see Table 10.1), which in this case is -19.4 to $+19.4$, rather than anything less than 18.5, the range for the one-tailed test.

Comparing the mean, m, with the acceptance level of $3.09s$ is clearly equivalent to comparing m/s with the acceptance level 3.09 for a standard normal deviate. For our polio data we can then compare $40.7/6.0 = 6.8$ with 3.09 for the 0.1% significance level and we again find that the mean differs significantly from zero.

7.2.1 *t* tests

Whenever we want to compare two means we proceed as in the example above, so let us write it out clearly, assuming that we are using a two-tailed test. To test if the mean of a set of numbers differs significantly from zero, divide the mean by the standard deviation: if the result is greater than 1.96, the mean differs significantly from zero at the 5% level; if it is greater than 2.58, the mean differs significantly from zero at the 1% level; if it is greater than 3.29, the mean differs significantly from zero at the 0.1% level; and so on. To add to our notation: the three most commonly used significance levels, 5%, 1% and 0.1%, are often indicated by *, ** and ***, respectively, while if a result is not significant at the 5% level we indicate this with the letters ns.

There is, however, an important correction we need to apply when we have only a few observations. Consider the following two numbers, chosen at random from a standard normal distribution: 0.218 and 0.332. We know that their true mean is 1 and their true variance is 1, since that is how they were chosen, and indeed both 0.218 and 0.332 lie well within the acceptance range of ± 1.96 for a 5% significance level. Suppose we didn't know in advance how they were chosen and wanted to test the hypothesis that their mean value differs from zero. The calculated mean, m, of 0.218 and 0.332 is 0.275 and the calculated population standard deviation is 0.081, so that the standard deviation of the mean, s_m, is 0.057 and $m/s_m = 4.8$. Since this is greater than 3.29, we would say that the mean differs significantly from zero at the 0.1% significance level and give it 3 stars! Now we know that this is **not** the case, since we deliberately chose the numbers from a normal distribution with a mean of zero. So where have we gone wrong? Our estimate of 0.275 for the mean is reasonable since we expect it to lie in the range ± 1. However, our estimated population standard deviation 0.081 is 12 times smaller than the true value of 1 and this is where our problem lies. To illustrate this further, I generated two more normally distributed numbers and obtained -0.969 and 0.582. For these two numbers the mean is -0.194 and the standard deviation is 0.775, so that $m/s_m = 0.25$, this time well within the acceptance range at the 5% significance level. In this second case the mean differs from zero by about the same amount as before but the estimate of the standard deviation is now much closer to the true value of 1. So the problem arises because we have had to use an estimate of the standard deviation instead of the true value, thereby introducing a further uncertainty. If we had used the true value of 1 for the standard deviation, we would have had 0.275 and 0.194 for the two estimates of m/s_m and in both cases would have accepted the hypothesis that the numbers came from a normal distribution with mean zero. Since we do not know the true value of the standard deviation, we need to allow for the inaccuracy inherent in our estimate of the standard deviation in setting the acceptance range for the ratio m/s_m.

The solution to this problem was found by Student (1908), who realized that m/s_m follows a normal distribution only if s_m is known exactly. He therefore set out to determine the distribution that is followed by m/s_m. Student was able to show that if m is the mean of n readings taken from a normal distribution whose true mean is μ, and if s_m is the estimated standard deviation of the mean, then

$$t = (m - \mu)/s_m \sim t_{n-1}. \qquad 7.2$$

In words, the t statistic in Equation 7.2 follows Student's t distribution with $n - 1$ degrees of freedom, the number used to estimate s_m. (Equation 7.2 is derived in the Appendix, section 7.10.1.) Of course, we do not in general know the true mean, μ, for if we did we would not be concerned with the estimate m. What we do is to make an hypothesis about μ and then test the difference $(m - \mu)/s_m$. In our example above, the hypothesis is that $\mu = 0$ and so we test m/s_m against t_{n-1}.

Let us look at our two pairs of numbers again. Since we are comparing two numbers, we have 2 degrees of freedom, but we have used up one of them when we used the mean to estimate the standard deviation. We therefore look up the t distribution with $2 - 1 = 1$ degree of freedom in Table 10.1. The acceptance range at a 5% significance level is $- 12.71$ to $+ 12.71$. Since our two ratios were 4.8 and 0.25, both of them now lie well within the acceptance range for the 5% significance level and we accept our null hypothesis that the two pairs of numbers come from a normal distribution with mean zero.

Since this is a crucial result let us restate it: when we divide the sum of the squares of the deviations by $n - 1$ when we calculate the standard deviation, subtracting the 1 allows for the fact that we have used the estimated mean, and not the true mean, and this gives us our best unbiased estimate of the standard deviation. The use of the t distribution rather than the normal distribution to test the ratio of the estimated mean to the estimated standard deviation allows for the fact that we have used the estimated rather than the true standard deviation, so that even our best estimate of s_m may not be very good. For very large samples the estimate of the standard deviation will be close to the true value and the t distribution tends to the normal distribution as the number of degrees of freedom increases as illustrated in Figure 5.6. We can therefore look on the t distribution as a correction to the normal distribution when the sample size is small.

To illustrate the use of the t distribution, Student (1908) analysed data on the effect on sleep of two drugs, hyoscyamine and hyoscine. The drugs were administered to ten patients in the Michigan Asylum for the insane at Kalamazoo and the results were repoted by Cushny and Peebles (1905). The amount of time each patient slept was measured for between 3 and 9 nights in each case with no drug and after the administration of 0.6 mg of hyoscyamine or hyoscine. The results are given in Table 7.2.

To test if hyoscyamine is soporific, we calculate the mean amount of sleep gained, as compared with the control (0.75 hours) and the standard deviation of the mean (0.57 hours). The ratio gives us a t value of 1.32. We have ten points and lose 1 degree of freedom in calculating the mean, so we are left with 9 degrees of freedom. Since we are interested in the drug only if it increases the amount of sleep, we use a one-tailed test. At the 5% significance level, t_9 is 1.83 (Table 10.1) so the effect of hyoscyamine is not significant. For hyoscine, the value of t is 3.68, which exceeds the value of t_9 at the 1% significance level, 2.82, and we conclude that hyoscine does increase the amount of sleep significantly and give it two stars.

So far we have tested each mean on its own to see if it differs from zero. We might also like to know if the two drugs differ significantly in thier effects. Now we want to see if the difference between two means differs from zero. The t test for the difference of two means is a straightforward extension of the above results. If we have n readings in each set of data so that the variance of the first mean is s_1^2/n, while that of the second is s_2^2/n, then the variance of the difference is $s^2 = s_1^2/n + s_2^2/n$.

Table 7.2 The number of hours for which patients slept with no drug, with hyoscyamine and with hyoscine. The increase in the amount of sleep gained with each drug as well as the difference in the amount of sleep gained with each drug are also given

Patient	Control	Hyoscyamine	Increase	Hyoscine	Increase	Difference
1	0.6	1.3	+ 0.7	2.5	+ 1.9	1.2
2	3.0	1.4	− 1.6	3.8	+ 0.8	2.4
3	4.7	4.5	− 0.2	5.8	+ 1.1	1.3
4	5.5	4.3	− 1.2	5.6	+ 0.1	1.3
5	6.2	6.1	− 0.1	6.1	− 0.1	0.0
6	3.2	6.6	+ 3.4	7.6	+ 4.4	1.0
7	2.5	6.2	+ 3.7	8.0	+ 5.5	1.8
8	2.8	3.6	+ 0.8	4.4	+ 1.6	0.8
9	1.1	1.1	+ 0.0	5.7	+ 4.6	4.6
10	2.9	4.9	+ 2.0	6.3	+ 3.4	1.4
m			0.75		2.33	1.58
s			1.79		2.00	1.23
s_m			0.57		0.63	0.39
t			1.32		3.68	4.06
			ns		**	**

We call s^2 the 'pooled' standard deviation since it combines the estimates from both sets of data. Dividing the difference in the two means by the standard deviation of the difference gives our t parameter with $2n - 2$ degrees of freedom. (This is strictly valid only if the 'true' variances, σ_1^2 and σ_2^2, are the same. If this is not the case, or if the number of measurements contributing to each of the means differs, an approximate test can be carried out in which we use the same t statistic but a rather more complicated expression for the number of degrees of freedom. This is discussed in the Appendix, section 7.10.2).

For Student's data on soporific drugs (Table 7.1), the two means are 0.74 ± 0.57, and 2.33 ± 0.63 hours so that

$$m_2 - m_1 = 1.58 \text{ hours}$$

$$s_{m_1 - m_2} = (0.57^2 + 0.63^2)^{1/2} = 0.85 \text{ hours} \qquad 7.3$$

$$t = 1.86.$$

Now we need to compare this with the acceptance range for a t distribution with $2 \times 10 - 2 = 18$ degrees of freedom. Since either drug could be better, we need to use a two-tailed test and since t_{18} at the 5% significance level is 2.10, the two drugs do **not** differ significantly in their effect.

7.2.2 Paired t tests

When analysing experimental data it is important to make full use of all of the available data. In comparing the two drugs whose effects are given in Table 7.2, we compared the mean effects of the two drugs, and if there had been 20 different patients this would have been all we could have done. However, each patient received both drugs and so we could instead calculate the difference between the effects of the two drugs on each patient separately and then see if the mean of the differences (rather than the difference of the means) is significant.

To illustrate the problem, the data of Table 7.2 are plotted in Fig. 7.2, from which it is clear that the individual patients vary greatly in their response to the drugs and that this tends to conceal any difference between the effects of the two drugs. However, in all cases the second drug induces at least as much sleep as the first and by looking at the difference between the effects of the two drugs on each patient, the last column in Table 7.2, we might hope to have a more sensitive test of the difference between the effects of the two drugs. The mean difference is 1.58 hours with a standard deviation of 0.39 hours, giving a value of t equal to 4.06. We have ten differences and have used up 1 degree of freedom in calculating the mean, leaving us with 9 degrees of freedom. Since either drug might be more effective, we need to use a two-tailed t test. At the 5% significance level, t_9 is equal to 3.25 (Table 10.1), so that the drugs do differ significantly at the 1% level.

The point to note is that when we compared means without taking into consideration the fact that both drugs were given to each patient, their effects were not significantly different, whereas when we included the additional information that each patient was given both drugs, their effects were significantly different. The second kind of test, in which we take the treatments in pairs, is called a paired t test. Obviously, if we had used 20 different patients and given each of them one of the drugs only,

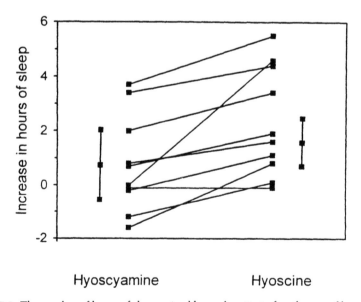

Figure 7.2 The number of hours of sleep gained by each patient after the use of hyoscyamine and hyoscine. The points with error bars give the mean effects of the two drugs $\pm$ 2.26s_m. The diagonal lines link the response of each patient to the two drugs.

instead of using ten patients each of whom received both drugs, we would not have been able to use a paired *t* test. Note in particular that in Table 7.2 the mean of the differences is precisely the same as the difference between the means. The increase in the sensitivity of the paired *t* test as compared with the unpaired *t* test is due to the reduction in the standard deviation from 0.85 hours, when we use the difference between the means, to 0.39 hours when we use the mean of the differences. By taking the differences between the responses to each drug separately for each patient, the paired *t* test calculates the variation due to the effect of the drugs after allowing for the average response of each patient.

7.3 NON-PARAMETRIC TESTS

When we compare two means using a *t* test, we assume that the individual measurements are normally distributed about their mean values. Although this assumption may be correct in the case of Student's data on the two drugs, we may have other sets of data for which the assumption of normality is not justified. We might even measure the response qualitatively so that we decide in each case which drug appears to be the better of the two without quantifying the effect in terms of hours of sleep. In such cases, we do not make precise assumptions about the underlying distribution or its parameters, and such tests are therefore called **non-parametric**.

Having carried out a parametric test on Student's data, let us see what happens if we carry out a non-parametric test. We will start with hyoscyamine but will now only use the fact that the drug increased the amount of sleep for five patients, produced no effect in one patient and decreased the amount of sleep in four patients. These observations are summarized in Table 7.3.

If hyoscyamine has no effect, we would expect an increase in the amount of sleep to be as likely as a decrease. (We are assuming that the distribution of the amount of sleep is symmetrical about the mean, but we are not assuming that the distribution is normal.) In other words

$$P(\text{increase}) = P(\text{decrease}) = 0.5. \qquad 7.4$$

Under our null hypothesis that the drug has no effect, the observations should come

Table 7.3 The number of patients for which hyoscyamine and hyoscine increased their amount of sleep, had no effect or decreased their amount of sleep. The last column gives positive, zero or negative differences between the amount of sleep gained when using each of the two drugs

	Hyoscyamine	*Hyoscine*	*Difference*
Increase	5	9	9
No effect	1	0	1
Decrease	4	1	0

Table 7.4 Binomial probabilities $P(k, n)$ and cumulative probabilities $C(k, n)$ for k successes in n equal to 9 and 10 trials

	Number of successes, k										
	0	1	2	3	4	5	6	7	8	9	10
$P(k, 9)$	0.002	0.018	0.070	0.164	0.246	0.246	0.164	0.070	0.018	0.002	●
$C(k, 9)$	0.002	0.020	0.090	0.254	0.500	0.746	0.910	0.980	0.998	1.000	●
$P(k, 10)$	0.001	0.010	0.044	0.117	0.205	0.246	0.205	0.117	0.044	0.010	0.001
$C(k, 10)$	0.001	0.011	0.055	0.172	0.377	0.623	0.828	0.954	0.989	0.999	1.000

from a binomial distribution with $p = 0.5$. Before we proceed, we need to decide what to do about the one patient who neither gained nor lost sleep when using the drug. Since the tie provides no evidence one way or the other, the usual procedure is simply to drop it from the analysis.

We can calculate the binomial probabilities for all possible outcomes of our experiment using Equation 5.2 and these are given in Table 7.4. We are interested in the drug only if it increases the amount of sleep. We see from Table 7.4 that if the drug has no effect, the probability of four or fewer successes in nine trials is 0.5, so that the probability of five or more is also equal to 0.5 and, on the basis of the available evidence, we conclude that hyoscyamine has no effect, which agrees with our previous conclusion.

With five successses and four failures out of nine trials, it should have been obvious that we would not have a significant effect. Let us therefore try the same test on hyoscine for which nine out of ten of the patients showed an increase in sleep. We see from Table 7.4 that the probability of eight or fewer successes in ten trials is 0.989, so that the probability of nine or more successes is only 0.011, and this outcome is very unlikely. We can therefore reject the null hypothesis at the 5% but not quite at the 1% level, which again agrees with our earlier conclusion.

We can now proceed to compare the effects of the two drugs. We now have nine increases and no decreases with one tie, which we ignore (Table 7.3). This time we need to use a two-tailed test since we do not know in advance which of the two drugs is the more efficacious. The probability of nine successes in nine trials is equal to the probability of zero successes in nine trials, which is 0.001. The probability of a result as extreme as this, in either direction, is therefore 0.002 and we can reject the null hypothesis and conclude that the two drugs differ significantly at the 1%, but not at the 0.1% level, once again in agreement with our previous conclusions.

The test described above is called the **sign test** because it is based on the signs of the differences between the effect of each drug and no drug and between the two drugs. Although our conclusions concerning the effects of the drugs were the same using the parametric t test and the non-parametric sign test, non-parametric tests are generally less powerful and this is the price that we pay for making fewer assumptions about the underlying distribution of the data. To confirm this we note that when we compared the two drugs using the two-tailed *t* test the significance level corresponding to the critical value of 4.08 with 18 degrees of freedom was 0.07%, whereas the significance level corresponding to zero or nine successes using the sign test was 0.2%. The *t* test was therefore significant at the 0.1% level while the sign test was significant only at the 1% level. Nevertheless, it is often the case that a suitable non-parametric test is almost as powerful as the corresponding parametric test and since non-parametric tests make fewer assumptions about the data, they also give more robust conclusions.

Although we will not discuss non-parametric tests further in this book, there are non-parametric tests corresponding to all of the standard parametric tests. Non-parametric tests are especially important in the social sciences where the data collected often take the form of answers to questionnaires and may be nominal categories that cannot be ranked, such as single, married, widowed or divorced, or ordinal categories that can be ranked but not on a quantitative scale, such as very bad, bad, good or excellent. You are also likely to encounter non-parametric tests in biology where an animal's behaviour may again be categorized but not on a quantitative scale. In these cases, tests may be carried out using only signs of differences, as in the examples given above, they may take advantage of ranking in the data, they may test runs of positive and negative values to ensure that series of data are random and not clumped. The important thing to bear in mind is that if the data are not measured on a quantitative scale or if you suspect that the distribution of the data does not follow a known, usually normal, distribution, you should use a non-parametric test. Good references for further reading are the books by Siegel and Castellan (1988) and Sprent (1990).

7.4 MULTIPLE RANGE TESTS

If we are comparing only two numbers, we proceed just as in the previous section. However, if we are comparing many numbers, we encounter a problem. Consider, for example, the housefly data of Table 3.1 in which the longest wing was 5.57 mm and the shortest 3.63 mm. The difference in the lengths of these two wings is 1.94 mm, while the sample standard deviation is 0.39 mm. The standard deviation of the difference is therefore $0.39 \times \sqrt{2}$, so that the *t* statistic is 3.51 with 99 degrees of freedom. If our hypothesis is that both measurements come from the same normal distribution with a population standard deviation of 0.39, we should reject the hypothesis at the 0.1% significance level!

It should be immediately clear that I have cheated by choosing the shortest and the longest wings: if I had chosen two wings at random, the comparison would have been valid. Of course, I might have picked out the longest and shortest wings even if the choice had been random, but the probability of that happening is only $(2/100) \times (1/99) = 2 \times 10^{-4}$, and we know that we cannot exclude the possibility of a Type I error completely. Nevertheless, if we do make several measurements there is no reason not to compare all of them; we simply have to be more cautious in the analysis.

There are a number of ways in which one can deal with the problem of multiple comparisons. Sokal and Rohlf (1981, p. 261) list ten different tests, along with the conditions under which you should use them. Here I consider the problem briefly.

If, before looking at the numbers, we decide to compare only two of them, we simply use the *t* test as before. However, we see that if we have 100 numbers and make all possible comparisons, we are bound to find spuriously significant differences when we compare the largest with the smallest numbers. To avoid this problem we can show (Appendix, section 7.10.3) that if we measure n random variables, and if the probability that any one of them is less than some value, say x, is α, then the probability that all of them are less than x is approximately α/n. Comparing numbers in pairs is a slightly different problem from that of testing a series of numbers separately, but this result suggests that if we make k comparisons and want an overall significance level of α, say, we should use the acceptance range corresponding to a significance level of α/k. For ten comparisons at the 1% significance level, we would use the acceptance range for the 0.1% significance level so that for the normal distribution our 1% acceptance range would be ± 3.29 instead of ± 2.58 (Table 10.1). Now if we make n measurements, we can make at most $k = n(n-1)/2$ comparisons. For this reason Fisher (Kendall *et al.*, 1983) suggested that if we want to make all possible comparisons between n numbers and ensure that the significance level for each comparison is at least α, we should use a significance level of $\alpha/[n(n-1)/2]$.

For our houseflies we can make up to $100 \times 99/2 = 4950$ comparisons. Testing the difference between the biggest and the smallest numbers at the 5% significance level, we than use the acceptance range for the $5/4950 = 0.001\%$ significance level with 99 degrees of freedom (the number of degrees of freedom in the calculation of the standard deviation). At this significance level $t_{99} = 4.5$ and the calculated t statistic for the difference between the lengths of the longest and shortest wings, 3.51, is well within the acceptance range. We accept the hypothesis that the difference is not significant at the 5% significance level.

We now have two extreme cases: the standard t test strictly applies only to a comparison between two numbers or two numbers chosen at random from a larger set of numbers and is sometimes called the least significant difference or LSD test. If we wish to compare more than two numbers, the acceptance range will be too narrow and it will produce too many Type I errors indicating significant differences where none exist. On the other hand, the modified t test applies only to a comparison between the largest and smallest of a set of numbers. If we wish to compare more than two means, the acceptance range will be too wide, it will produce too many Type II errors and fail to pick up significant differences.

To illustrate this, Table 7.5 shows the critical values that define the acceptance ranges at the 5% significance level for the LSD and modified t test (the first and third number in each triplet) for various numbers of measurements and degrees of freedom. If we have only two measurements to compare, the critical values are the same for both tests, as we would expect. But as we increase the number of measurements that we wish to compare, the modified t test becomes progressively more conservative by comparison with the LSD test since a bigger value of t is required before we can reject the null hypothesis.

Table 7.5 Critical values at the 5% significance level for the difference between the biggest and the smallest of a set of normally distributed numbers. The number of comparisons that are made is n and the number of degrees of freedom used to calculate the standard deviation is f. The first number in each triplet is based on the LSD or standard t test and is $\sqrt{2}t_f(0.975)$. The second number in each triplet is based on the Studentized range statistic $q(n, f)$. The third number in each triplet is based on the modified t test and is $\sqrt{2}t_f(1 - 0.025/k)$ where $k = n(n-1)/2$

Degrees of	Number of measurements, n			
freedom, f	2	5	10	20
	18.0	18.0	18.0	18.0
1	18.0	37.2	49.1	59.6
	18.0	180	810	3421
	3.64	3.64	3.64	3.64
5	3.64	5.67	6.99	8.21
	3.64	6.74	9.49	12.9
	3.15	3.15	3.15	3.15
10	3.15	3.58	4.52	5.50
	3.15	4.66	5.60	6.47
	2.77	2.77	2.77	2.77
20	2.77	3.86	4.39	5.01
	2.77	3.97	4.61	5.16

The modified t test is only approximate, even when testing the difference between the largest and the smallest numbers from a set of numbers. However, we can calculate the so-called Studentized range statistic, $q(n, f)$, which is the actual distribution of the difference between the largest and the smallest of two numbers taken from a set of n normally distributed numbers, where f is the number of degrees of freedom used to calculate the standard deviation. The test based on the Studentized range statistic is called Tukey's honestly significant difference, or HSD, test (Hays, 1988, p. 420) and critical values are given by Hays (1988, p. 949). In Table 7.5 the middle number in each triplet gives the critical value for the HSD test and this lies between the critical values for the LSD test and the modified t test. Provided the number of degrees of freedom is greater than the number of points that we wish to compare, the three tests do not differ greatly. When the number of degrees of freedom is small and the number of points to be compared is large, they do differ greatly but it would be a bad experiment that compared, say, 20 measurements on the basis of a standard error calculated with only 1 degree of freedom.

There are two important points to be made here. If two numbers do not differ significantly using the LSD test they will not differ significantly on any test. Equally, if two numbers do differ significantly using the modified t test they will differ significantly on any other test. It is only when we find marginally significant differences using the LSD test or marginally non-significant differences using the modified *t* test that we need to exercise caution.

The second point is that even the HSD test based on the Studentized range statistic is exact only when comparing the largest and the smallest of two numbers from a given set of numbers. We would expect numbers that lie close together in the ordered distribution to differ by less. To allow for this, we can use a variation on the HSD test called the Student-Newman-Keuls test (Hays, 1988, p. 420), in which we first compare the largest and the smallest of *n* numbers using the Studentized range statistic as described above. If, and only if, they differ significantly, we then compare the two numbers that are one step closer together in the ordered sequence using the Studentized range statistic based on $k - 1$ comparisons. We then proceed inwards in this way, stopping as soon as we find the first non-significant difference.

7.4.1 Prussian army officers

Let us look again at the deaths of Prussian army officers in Table 4.1. We still feel uneasy about army corps number 9 in which 24 officers were kicked to death by their horses. Table 7.6 gives the number of officers kicked to death in each corps, summed over all 20 years. To see if army corps number 9 did have significantly more deaths than the others, our null hypothesis is

$$H: \text{The number of deaths in each army corps}$$

$$\text{follows the same Poisson distribution.} \qquad 7.5$$

The mean number of army officers kicked to death in all of the corps is 12. The probability that a number, chosen from a Poisson distribution with a mean of 12, is greater than or equal to 24 is 0.16%. The number of deaths in army corps 9 therefore appears to be significant at the 1% significance level. However, we have deliberately chosen the greatest value and since we have ten numbers we are effectively making ten comparisons. To test the largest number at a significance level equal to *P*, we need to use the critical value for a significance level of $P/10$. The probability that the largest of ten numbers chosen from our Poisson distribution is greater than 24 is therefore equal to $0.16 \times 10 = 1.6\%$, so that we can reject Hypothesis 7.5 at the 5% level but not at the 1% level.

Once again, we are left to decide what it means. There may well have been a problem with safety in army corps number 9. However, army corps number 9 might have been bigger than the others, had more cavalry than the others or simply been more active than the others. The statistics raise a question: we have to provide the answer.

Table 7.6 The number of army officers kicked to death in each of ten army corps between 1875 and 1894

Corps	1	2	3	4	5	6	7	8	9	10
Deaths	12	12	8	11	12	7	13	14	24	8

7.5 VARIANCE RATIOS

So far we have concentrated on comparisons between means and we have assumed that the variance of the sets of numbers used to calculate the means are the same. But suppose we want to compare two variances. We saw in section 5.4.3 that since each variance follows a χ^2 distribution, the ratio of two variances will follow an F distribution so that

$$V_1/V_2 \sim F_{n_1-1,n_2-1} \qquad 7.6$$

where $n_1 - 1$ and $n_2 - 1$ are the number of degrees of freedom associated with V_1 and V_2, respectively (Appendix section 7.10.4).

We can illustrate this with data collected to investigate the basis of speciation. Mayr (1942) suggested that a new population that starts from a small group of founders will be genetically atypical of the parent population and thus more likely to give rise to a new species than a new population starting from a large group of founders. To test this hypothesis, Dobzhansky and Pavlovsky (1957) took 20 populations of fruit flies: ten having 5000 individuals each and ten having 20 individuals each. At the start of the experiment, 50% of the individuals in each population were homozygous for a gene P, 50% were heterozygous. At the end of the experiment, after 18 generations, the proportions of flies homozygous for the gene P were as given in Table 7.7. The mean proportion of homozygous flies in the large populations is 0.26 with variance 0.0032, while for the small populations the mean is 0.32 with variance 0.0127. The ratio of the two variances is 3.96 and we need to compare this with an F distribution with 9 and 9 degrees of freedom. From Tables 10.3 and 10.4 we see that the variances differ significantly at the 5% but not at the 1% level, giving support to Mayr's theory.

7.6 CONFIDENCE LIMITS

In the analyses developed so far we have (i) made an hypothesis, H, about the data (e.g. two means are the same), (ii) made a series of measurements from which to calculate a test parameter, T, (e.g., the difference between the two means divided by the standard deviation), (iii) determined a range, R, within which T should lie if H is true (e.g. 95% acceptance range for a normally distributed variable), (iv) accepted H if T is contained in R; rejected H if T is not contained in R.

It is important to be clear as to what we mean when we accept or reject H. When we accept an hypothesis, we are not saying that it is necessarily true but only that we have failed to show that it is false; more accurate data may well lead us to reject the hypothesis. Although we accepted the null hypothesis that hyoscyamine does

Table 7.7 The proportions of flies bred from large and small initial populations that were homozygous for the gene P after 18 generations

Large	0.187	0.202	0.208	0.216	0.244	0.276	0.306	0.312	0.324	0.338
Small	0.153	0.172	0.208	0.306	0.321	0.333	0.440	0.306	0.450	0.462

not increase the amount of sleep significantly (section 7.2.1), more extensive data might reveal a significant difference and lead us to reject the null hypothesis.

We are on firmer ground when we reject, rather than accept, an hypothesis, and we can be confident that the hypothesis does not hold. In the trial of the polio vaccine, we were able to show that the vaccine produces a statistically significant reduction in the number of children contracting polio. Nevertheless, the data in themselves do not tell us why the null hypothesis is false or what would be a reasonable alternative hypothesis. In the study of children in Saxony, we decisively rejected the hypothesis that the data follow a binomial distribution in which the probability of having a boy or a girl is independent of the sex of its older siblings. However, the observed numbers deviate only slightly from the binomial prediction and any further hypothesis should take the binomial model as a starting point.

Considering hypotheses in this manner is both rigorous and sound, but rigour alone leads to pedantry, not progress. For example, in our study of the effect of hyoscine on sleep we found that the average sleep gained was 2.33 ± 0.63 hours so that m/s was about 4 and we concluded that the number of hours of sleep gained differed significantly from zero and that the hyoscine does induce sleep. But we should be able to say more than this. For example, I think we agree that it is unlikely that hyoscine will increase the amount of sleep by, say, 10 hours or that it decreases the amount of sleep. What we want is an idea of the limits within which we can be confident that the 'true' value lies. Let us see if we can define confidence limits for our estimates of various parameters so that we can regard the sample mean, for example, as our 'best' estimate of the underlying population mean and use the standard deviation as an indication of the range within which the true mean should lie.

Setting this out formally, suppose that we make a series of n measurements on a variable that is normally distributed with true mean μ, estimated mean m and estimated standard deviation of the mean equal to s_m. Now we have seen (Equation 7.2) that

$$t = (m - \mu)/s_m \qquad\qquad 7.7$$

follows a t distribution with $n - 1$ degrees of freedom, so that if t^* is the upper 97.5% point of the distribution,

$$P(-t^* < t < t^*) = 0.95, \qquad\qquad 7.8$$

and there is a 95% probability that t lies in the range $\pm t^*$. We use the critical value for a two-tailed test because we do not know if m is less than or greater than μ. Now if $t < t^*$, then $(m - \mu)/s_m < t^*$ and $\mu > m - t^*s_m$. Similarly, if $-t^* < t$, then $\mu < m + t^*s_m$, so that

$$P(m - t^*s_m < \mu < m + t^*s_m) = 0.95, \qquad\qquad 7.9$$

which tells us that there is a 95% probability that the range $m \pm t^*s_m$ includes the true mean. We can assert that the amount of sleep gained after taking hyoscine is 2.33 ± 1.43 hours, that is between 0.90 and 3.76 hours, with 95% confidence.

In our study of the polio vaccine (section 7.2), we found that the difference between the number of vaccinated and unvaccinated children who contracted polio was 40.7 ± 6.0, and we are able to reject the hypothesis that the vaccine has no effect at the 0.1% significance level. We can now use our concept of confidence limits to say instead that the reduction in the number of children per 100 000 who will develop polio when given the vaccine is $40.7 \pm 1.96 \times 6.0$, that is, in the range 29 to 52 per 100 000, with 95% confidence. We can also say that this number lies between 21 and 60 per 100 000, with 99.9% confidence.

Expressing our results in terms of confidence limits appears to be much stronger than merely accepting or rejecting an hypothesis. However, we have to be careful about the interpretation of confidence limits. We cannot strictly conclude from Equation 7.9 that the true mean lies within the defined confidence limits with the calculated probability because we know that it either lies within the limits with certainty or outside the limits with certainty. What we can do is simply assert that it lies within the given interval and we will then be right in 95% of cases and wrong in 5% of cases, for 95% confidence limits. Alternatively, we can note that we are effectively using our confidence limits to test all possible hypotheses about the value of the mean and the confidence limits then correspond to those values we would accept while excluding all those values we would reject. Regarding the polio study, where we found a difference in the number of children, per 100 000, contracting polio of between 29 and 52, with 95% confidence, we are saying that we would reject the hypothesis that the true difference is 0, 1, 2,..., 27, 28 or 53, 54,...at the 5% significance level but we would accept the hypothesis that the true difference is 29, 30,..., 51, 52 at the 5% significance level. (Bulmer, 1979, provides an excellent discussion of this topic but see also Kendall and Stuart, 1983.)

7.6.1 The importance of 'three'

The discussion of confidence limits leads to an important observation. Whenever you carry out an experiment or make an observation, you should always take at least three readings. One reading gives us an estimate of the mean but no indication of the dispersion. Two readings enable us to calculate the standard deviation, and a 95% confidence interval for the mean is then

$$m \pm t_1(0.975)s/\sqrt{1} = m \pm 12.7s. \qquad 7.10$$

Two repeats is thus the absolute minimum number of readings you should take. However, with three repeats the confidence interval for the mean is

$$m \pm t_2(0.975)s/\sqrt{2} = m \pm 4.3s/\sqrt{2} = m \pm 3.0s. \qquad 7.11$$

With 50% more effort, we have increased the precision of our estimate by a factor of 4! If we take still more readings, we improve the accuracy further so that for four readings we have $m \pm 1.6s$ and for five readings we have $m \pm 1.1s$. None of these gains, however, is as striking as that made in going from two to three readings.

7.7 SEVERAL FACTORS

In many experiments the response variable that we measure depends on several different factors and we want to examine the effects of each factor separately as well

Table 7.8 The weight (in grams) gained by male and female rats fed on diets containing fresh and rancid lard

	Fresh	Rancid
Male	171	108
	172	89
	172	69
Female	153	85
	109	64
	160	82

as the interactions between them. We can illustrate this using data from an experiment to discover whether rancid lard has a lower food value than fresh lard. Because rancid lard contains strongly oxidizing peroxides and vitamin A is readily oxidized, Powick (1925) felt that the consumption of rancid lard might cause rats to suffer from a deficiency of vitamin A. Three young male and three young female rats were fed on a diet that included fresh lard and the same number of rats were fed on an equivalent diet but using rancid instead of fresh lard. To minimize the effects of variables other than those arising from the difference in the diets, the rats were selected for uniformity as to age, weight and general health. The rats each weighed between 39 and 52 g at the start of the experiment. Table 7.8 gives the weight gained by each rat after 73 days.

In this experiment we will treat each group of three rats as repeats of the measurement of the effect of sex and lard freshness on the weight gained by the rats. Everything we need to know about the data in Table 7.8 is contained in the mean and the standard deviation of each set of three numbers and these are given in Table 7.9. We want to use these data to answer the following questions:

- Does the freshness of the lard affect the weight gain of the rats?
- Does the sex of the rats affect the weight gain of the rats?
- Does the effect of lard freshness on the weight gains depend on the sex of the rats? Or, putting it the other way round, does the effect of sex on the weight gains depend on the freshness of the lard?

The first question is then: does the freshness of the lard affect the weight gained by the rats? We could answer this simply by comparing the average weight gained

Table 7.9 The mean and standard deviation of the weight (in grams) gained by male and female rats fed on a diet containing fresh and rancid lard

	Fresh	Rancid
Male	171.7 ± 0.33	88.7 ± 11.25
Female	140.7 ± 15.96	77.0 ± 6.56

by all of the rats that were fed on fresh lard with the average weight gained by all of the rats that were fed on rancid lard. This is precisely what we would do if we did not know the sex of the rats and we would use an unpaired t test to do it. However, we have seen that, whenever possible, we should do a paired t test as this is always more powerful. Unfortunately, in the example of our rats we are unable to pair them directly since each rat was fed on rancid lard or fresh lard but not both. However, we can pair the rats by sex so that we consider the effect of lard freshness on weight gain for male rats and the effect of lard freshness on weight gain for female rats separately, and then calculate the effect of freshness averaged over the effects on males and females.

From Table 7.10 we see that the male rats gain 83.0 g more weight when the lard is fresh rather than rancid and that the standard deviation of this difference is $\pm$ 11.3 g (remembering that the variance of a difference is the sum of the variances). Similarly, the female rats gain 63.7 $\pm$ 17.3 g more when the lard is fresh rather than rancid. In both cases the rats gain more weight when fed on fresh rather than rancid lard and the average increase is 73.0 $\pm$ 10.3 g.

Before we perform t tests to determine the effect of freshness and sex on the weight gain of the rats we need to decide how many degrees of freedom to use. Since each mean in Table 7.9 is calculated using 3 points, each standard deviation in that table has 2 degrees of freedom. When we compare the weight gained by male rats eating fresh lard and rancid lard we therefore have $2 + 2 = 4$ degrees of freedom, and of course the same applies to the comparison for female rats as well as to comparisons between the sexes fed on fresh lard and on rancid lard.

For the male rats then our t statistic for the effect of freshness on weight gain is $83.0/11.3 = 7.4$ and using a two-tailed t test this is significant at the 1% level but not at the 0.1% level (Table 10.1) and has two stars. For the female rats we see from

---

Table 7.10 The effect of sex and freshness on the weight gain of rats. F — fresh lard, R — rancid lard, m — male rats, f — female rats. The first line, for example, gives the average difference between the weight gained when eating fresh as opposed to rancid lard for male rats

Lard freshness

$$\langle F - R | m \rangle = 83.0 \pm 11.3 \qquad t_4 = 7.4^{**}$$
$$\langle F - R | f \rangle = 63.7 \pm 17.3 \qquad t_4 = 3.7^{*}$$
$$\langle \langle F - R | m \rangle + \langle F - R | f \rangle \rangle = 73.3 \pm 10.3 \qquad t_8 = 7.1^{***}$$

Sex

$$\langle m - f | F \rangle = 31.0 \pm 16.0 \qquad t_4 = 1.9 \, ns$$
$$\langle m - f | R \rangle = 11.7 \pm 13.0 \qquad t_4 = 0.9 \, ns$$
$$\langle \langle m - f | F \rangle + \langle m - f | R \rangle \rangle = 21.3 \pm 10.3 \qquad t_8 = 2.1 \, ns$$

Effects of sex on effect of freshness
$$\langle F - R | m \rangle - \langle F - R | f \rangle = 19.3 \pm 20.6 \qquad t_8 = 0.94 \, ns$$

Effect of freshness on effect of sex
$$\langle m - f | F \rangle - \langle m - f | R \rangle = 19.3 \pm 20.6 \qquad t_8 = 0.94 \, ns$$

the data summarized in Table 7.10 that the effect of freshness on weight gain is significant only at the 5% level and has one star.

To determine the overall effect of lard freshness, we average the effect for males and females and this gives us an average weight gain when the lard is fresh of 73.3 ± 10.3 g. We are now calculating the average of two numbers each of which has 4 degrees of freedom so that the number of degrees of freedom for this comparison is 8. Our t statistic is $73.3/10.3 = 7.12$ and this is significant at the 0.1% level and has three stars. (If we had not distinguished between the male and female rats, the mean increase in consumption would have been the same, i.e. 73.3 g, but the standard deviation would have been 26.4 g instead of 10.3 g. Using a two-tailed t test with 10 degrees of freedom, the value of $t = 2.78$ would have been significant at the 5% but not at the 0.1% level. Although the effect of the sex of the rats is not in itself significant, by allowing for the small effect of sex, we obtain a more sensitive test of the effect of lard freshness.)

We can now perform the same set of calculations for the effect of sex on the consumption of lard. The results of these calculations are also summarized in Table 7.10, from which we see that in fact none of them is significant at the 5% level and all receive an *ns*.

The results of these calculations show that the weight gained by the rats is influenced by the freshness of the lard but not by the sex of the rats. But it is also possible that the sex of rats interacts with the effect of the freshness of the lard. The concept of interactions in important but rather subtle. To illustrate what is meant by an interaction, suppose that male and female rats both gained 100 g more when the lard was fresh rather than rancid. The difference in weight gain due to lard freshness would then be independent of the sex of the rats. But if the male rats gained 100 g more when the lard was fresh while the females gained 100 g less when the lard was fresh, we would conclude that the effect of freshness does depend on the sex of the rats and we call this effect an **interaction between freshness of the lard and sex of the rats**.

To calculate the interaction between the freshness of the lard and the sex of the rats, which we write lard × sex, we take the effect of lard freshness on the weight gain of male rats, which is 83.0 ± 11.3 g, and compare this with the effect of lard freshness on the weight gain of female rats which is 63.7 ± 17.3 g. The difference between the effect of freshness for male rats and the effect of freshness for female rats is then 19.3 ± 20.6 g. Now we are comparing two numbers each with 4 degrees of freedom so that the number of degrees of freedom in the interaction term is 8 and our t statistic is $19.3/20.6 = 0.94$, which is not significant.

The sex of the rats does not affect the change in consumption due to the freshness of the lard: is it possible that the freshness of the lard affects the change in consumption due to the sex of the rats? The two numbers we now wish to compare are, from Table 7.8, 31.0 ± 16.0 g and 11.77 ± 13.1 g, from which we find a difference of 19.3 ± 20.6 g, exactly the same as before, so that the interaction term is the same whichever way we calculate it.

7.7.1 Contrasts

In the previous section we compared the effects of fresh and rancid lard, male and female rats, fresh lard consumed by male rats with rancid lard consumed by male rats, and so on. Altogether we made eight comparisons. However, there must be a limit to the number of independent comparisons or **contrasts** one can make (Winer, 1971). For example, if we have three numbers—7, 5 and 3, say—the first is greater than the second and the second is greater than the third. But this means that the first must also be greater than the third so that we can only make two independent comparisons.

The maximum number of independent pair-wise comparisons that we can make is one less than the number of independent means that we are comparing. In our example of the consumption of lard by rats, we have four independent means (Table 7.9) and we can therefore make at most three independent comparisons. The choice of which comparisons to make depends on what we want to learn from the data. We usually choose to compare each 'main effect' (lard freshness and sex) averaged over the other (sex and lard freshness) and then their interaction (the effect of sex on the change in consumption brought about by lard freshness or vice versa). If we choose to make three comparisons, say, we should divide our significance levels by 3 as we discussed in section 7.4 for multiple range tests. In this example, both sex and the interaction remain non-significant, as they must, while lard freshness is still significant at the 0.1% level.

7.8 EXPERIMENTAL DESIGN

The data we collect are only as good as the design of our experiments. In her experiment on rats, for example, Powick was careful to select the rats for uniformity of age, weight and general health so that these factors would not affect the outcome of her experiment. In the trials of the polio vaccine, we cannot give the placebo and the vaccine to the same children and so we have to find children who are likely to respond in the same way. We would not, for example, give the placebo to boys and the vaccination to girls because the efficacy of the vaccine might depend on the sex of the child. Indeed, we would want to give both the placebo and the vaccine to children from each state, each city and each school in the study area. However, we also have a moral dilemma; we would not be testing the vaccine if we did not believe that it was beneficial, so how shall we decide who should be given it and who should not? One way to do this is to use a random number generator to allocate the placebo and the vaccine within each sex in each school. This way each child has an equal chance of having the vaccine. Finally, to ensure that there is no unconscious bias, neither the doctors administering the vaccinations nor the doctors who eventually are to diagnose the patients for polio should know which injections contain the placebo and which contain the drug.

To illustrate the consequences of not carrying out well-designed trials, Rensberger

(1983) describes an operation, used for many years to treat people with angina, in which the mammary arteries are tied off. Doctors and patients were happy with the results. When a proper trial was eventually conducted using a control group, the control group, whose arteries were not tied off, fared better than those whose arteries were tied off. The operation was quietly dropped.

Another example of how not to proceed can be found in a series of trials in Kenya in 1988 to test the use of low-dose interferon administered orally for the treatment of AIDS patients (Pacala, 1990). The tests were carried out without a control group but the scientists doing the tests claimed striking improvements in the treated patients. The World Health Organization (WHO) then decided to run trials of their own. These trials were also carried out without controls in the hope that the Kenyan findings 'could be readily and immediately confirmed'. Eventually, the WHO concluded that though some patients appeared to improve, it was impossible to say whether this was due to the treatment or to the patient's own raised expectations. Moreover, the interferon used was not prepared in a consistent form. The WHO report concluded that properly controlled clinical trials should be 'undertaken in accordance with an appropriate experimental design in which all relevant variables are monitored' (Pacala, 1990). The attempt to cut corners simply meant that two years after the first trial no useful evidence as to the efficacy of the drug had been obtained. Now a Canadian team have begun a properly controlled trial to see if the issue can be decided one way or the other (Brown, 1991).

7.8.1 Crossed and nested designs

Experiments are often much more complicated than those we have considered so far. In Chapter 8 we will discuss an experiment designed to compare different traps for catching tsetse flies in which the catch depends on the trap, the site in which the trap is placed and the day on which the flies are caught. We will discuss an experiment to determine the amount of calcium in plants which will be seen to depend on the individual plants, the leaves within each plant and the samples taken from each leaf. In Chapter 9 we will consider the growth of Japanese larch trees and this will depend on the amount of nitrogen, phosphorus, potassium and residual ash. We need to think carefully about the design of such experiments to ensure that we can assess the effect of each variable after allowing for the effects of the other variables and, where possible, to assess the effect of each variable independently of all other variables.

The design of experiments warrants an entire book in itself and Mead and Curnow (1983) provide a good introduction. However, the most complicated experimental designs can be analysed in terms of crossed and nested factors. This is best explained using examples.

In the trial of the polio vaccine (section 7.2), the children were nested within the treatment since one set of children received the vaccine while another set received the placebo. We were unable to do a paired comparison since each child could not be given both the vaccine and the placebo.

When we used Student's data to compare two soporific drugs using an unpaired t test (section 7.2.1), the patients were nested within the drugs. However, when we considered the same data using a paired t test (section 7.2.2), the patients were crossed with drugs since each patient received both drugs and each drug was given to each patient. Although the difference between the mean effects of the drugs (unpaired t test, nested factors) was the same as the mean of the differences (paired t test, crossed factors) the latter analysis gave a more sensitive test because it separated out the variation due to patients leaving a smaller residual error.

In Powick's experiments on rats, sex was crossed with lard freshness since male and female rats were fed on fresh and rancid lard. We were able to assess the effects of the lard freshness after allowing for the effects of the sex of the rats. The individual rats, however, were nested within sex and lard freshness. The error term in our assessment of the effect of lard freshness on weight gain was smaller when we allowed for the effect of the sex of the rats than it was when we did not (section 7.7).

In Chapter 8 we will analyse experiments involving crossed and nested designs in more detail. The important thing to remember is that the analysis, and in particular our assessment of the errors, will depend on whether or not factors are crossed or nested.

7.8.2 Sample-size determination

As we increase the number of measurements that we make, the standard deviation of the mean decreases and we are more likely to be able to detect differences between pairs of means. Can we decide, in advance, how many measurements we should make in a particular experiment?

Consider again Student's data on soporific drugs (Table 7.2). Although hyoscyamine increased the mean amount of sleep in ten patients by 0.75 hours, the increase was not significant at the 5% level and may therefore have been due to chance. Suppose that we decided to conduct further tests on the drug. How many patients should we use to decide if the drug has an effect?

We need to know two things: the intrinsic variation in the data, that is, the population standard deviation, and the smallest effect that we wish to detect. The larger the intrinsic variation in the data and the smaller the effect we wish to detect, the more patients we will need to use.

If we let μ be the increase in the amount of sleep induced by hyoscyamine, our null hypothesis is

$$H_0: \mu = 0 \text{ hours.} \tag{7.12}$$

Now suppose that we are interested in the drug only if it increases the amount of sleep by at least one hour. Then our alternative hypothesis is

$$H_1: \mu = 1 \text{ hour.} \tag{7.13}$$

We now let n be the number of measurements that we will make and let x be the critical value of the increase in the amount of sleep. If the mean amount of sleep gained is less than x we will accept H_0; if the mean amount of sleep gained is greater than x, we will reject H_0 and implicitly accept H_1. To ensure that the probability of making a Type I error is less than, say, 5% we require

$$x \geqslant 1.65s/\sqrt{n}, \qquad 7.14$$

where s is the population standard deviation and assuming that the number of measurements is sufficiently large to allow us to use the normal approximation for t. But we might also want to keep the probability of making a Type II error small. Since the consequence of making a Type II error (deciding that an effective drug has no effect) may be less serious than the consequences of making a Type I error (deciding that an ineffective drug does have an effect), we might decide to keep the probability of making a Type II error less than 10%. Then we require

$$1 - x \geqslant 1.28s/\sqrt{n}, \qquad 7.15$$

since $1 - x$ is the difference between the mean and the critical value if our alternative hypothesis is true. Rearranging Equation 7.15 gives

$$x \leqslant 1 - 1.28s/\sqrt{n}. \qquad 7.16$$

The smallest value of n that ensures that both inequalities 7.14 and 7.16 hold will be the value for which they become equalities, so that

$$1.65s/\sqrt{n} = x = 1 - 1.28s/\sqrt{n}. \qquad 7.17$$

Our best available estimate of the population standard deviation is the value we already have, namely 1.8 hours. Solving Equation 7.17 for n we find that n must be at least 28. In other words, to decide if the drug induces at least one hour of extra sleep while keeping the probability of making a Type I error less than 5% and the probability of making a Type II error less than 10%, we need to use at least 28 patients. If the effect of the drug is to increase the amount of sleep by more than one hour, the probability of making a Type II error will be reduced so that 28 pairs of patients gives us a conservative estimate of the number of patients that we need to use. The critical value for our test is obtained from Equation 7.17 with $s = 1.8$ hours and $n = 28$ patients which gives $x = 0.56$ hours or 34 minutes. If the mean amount of sleep gained using hyoscyamine is less than 34 minutes we accept the null hypothesis and conclude that the drug does not have a significant effect, otherwise we reject the null hypothesis and conclude that the drug does have a significant effect.

We can apply similar arguments to our study of the polio vaccine. We want to determine the number of children we should use. Let us assume that before the trial we already know that the probability that an unvaccinated child will develop polio is $p_0 = 0.00057$, the value obtained from the unvaccinated children in our clinical trial (Table 1.1) since we will already have some knowledge of the incidence of the disease. Then if n children receive the placebo, we expect $np_0 \pm \sqrt{np_0}$ children to develop polio since the number of sick children should follow a Poisson distribution.

Since the consequences of making a Type I error, and concluding that an ineffective vaccine is effective, may be quite serious, we might decide to keep the probability of making a Type I error below 1%. The critical value, x, for our test must therefore be

$$x \leqslant np_0 - 2.33\sqrt{np_0} \qquad\qquad 7.18$$

Under our alternative hypothesis, we expect $np_1 \pm \sqrt{np_1}$ children to develop polio. We might now decide that we are interested in the vaccine only if it reduces the probability of a child developing polio by at least 50%, so that $p_1 \leqslant 0.00029$. Since making a Type II error may be less serious than making a Type I error, we might choose to keep the probability of making a Type II error to less than 5%. Then the critical value, x, for our test must also satisfy

$$x \geqslant np_1 + 1.65\sqrt{np_1}. \qquad\qquad 7.19$$

Once again the smallest acceptable value of n will be that for which both inequalities become equalities so that

$$np_0 - 2.33\sqrt{np_0} = np_1 + 1.65\sqrt{np_1}. \qquad\qquad 7.20$$

Solving Equation 7.20 for n gives

$$\sqrt{n} = (1.65\sqrt{p_1} + 2.33\sqrt{p_0})/(p_0 - p_1) = 289, \qquad\qquad 7.21$$

so that n is equal to 83 000. In other words, we see that while using about 200 000 children in each group might seem excessive, it is in fact only slightly conservative if we are to obtain a reliable test of the vaccine. Further discussions of sample size determination are given by Healy (1981) and Winer (1971, p. 30).

7.9 SUMMARY

When applying statistical tests to problems in biology, we want to know if a particular test statistic that we have calculated from our data could reasonably have come from a certain distribution. To do this we set up an acceptance range and a rejection range. However, the acceptance range might correspond to the upper part of the distribution if it is large values that we treat with suspicion or it might come from the lower part of this distribution if it is small values that we treat with suspicion. In both cases we will use a one-tailed test but take opposite ends of the distribution for our rejection range. It might be that we regard both large and small values with suspicion, in which case we will use both ends of the distribution to define our rejection range. Effectively we are choosing the rejection range in such a way as to maximize the power of our test in relation to the alternative hypothesis that we will accept if the null hypothesis is rejected.

To decide if two means differ significantly, we compare their difference and the standard deviation of their difference. The standard deviation is only an estimate of the true underlying standard deviation and it was once thought that this meant that tests could not be performed on small samples for which the estimated standard deviation is quite inaccurate. The importance of Student's discovery was that we do not in fact need to know the true standard deviations provided we use the *t* distribution rather than the normal distribution, for this allows for the inaccuracies associated with our estimate of the standard deviation.

There is, however, a second trap for the unwary. The *t* test applies to a comparison of two means. If we make 100 such comparisons and set a significance level of 5% about five comparisons will appear to be significant even if all the true means are exactly the same. We therefore need to use a more conservative acceptance range when we make many comparisons. Unfortunately, there is no universally best way to do this and I have outlined some of the ways that are commonly used. Of course if the significance level of the test is much less than 0.1% the test will almost certainly be significant however we do it. Equally, if the significance level of the test is much greater than 5% the test will almost certainly not be significant however we do it. You only need to tread carefully if the calculated value of the significance level is close to the range from 5% to 0.1% or if the number of comparisons that you make is very large.

The idea of paired and unpaired tests or crossed and nested designs is important because a paired test or a crossed design allows us to estimate the effect of one factor after allowing for the effects of another. Of course if we put in the wrong biology, and assume for example that the patients are paired when they are not, we will not have a meaningful answer. The biology is also important when deciding between parametric and non-parameteric tests since every test that we carry out involves a set of assumptions about the underlying distribution of the data. If we have reason to believe that these assumptions do not hold, it will be better to use a less powerful test that does not rely on these assumptions. As long as you understand the biology well you should be able to decide which test is appropriate.

The concept of confidence intervals provides us with a further powerful analytical tool because we do not only want to say a certain measurement is significant, but also want to set limits on the range of values within which we believe the true value lies. As long as we are clear as to how to interpret the confidence interval, there should be no problems in using it.

This and the previous chapter contain many of the most important concepts in the statistical analysis of data, but so far we have restricted our attention to fairly simple experimental designs. As the design of the experiment becomes more complicated, the number of comparisons that we can make, including the effects of interactions among the various factors in our experiments, increases alarmingly. In the next chapter we will discuss analysis of variance, or ANOVA, which provides us with a systematic way of analysing the results of more complicated experimental designs.

Finally, we have seen that although the power of a test is seldom used in reporting the results of our experiments, it provides an important tool in helping us to decide on how big a sample we need for a particular experiment.

7.10 APPENDIX

7.10.1 Derivation of student's t test

Suppose we have n measurements, x_i, $i = 1, 2, \ldots, n$ from a normal distribution with true mean μ and true variance σ^2 so that

$$x_i \sim N(\mu, \sigma^2). \tag{7.22}$$

Then

$$(x_i - \mu)/\sigma \sim N(0, 1) \tag{7.23}$$

and

$$\Sigma(x_i - \mu)^2/\sigma^2 \sim \chi_n^2. \tag{7.24}$$

When we calculate the standard deviation, we use the estimated mean m, rather than the true mean, μ, and this reduces the number of degrees of freedom in the sum by 1 so that

$$\Sigma(x_i - m)^2/\sigma^2 \sim \chi_{n-1}^2 \tag{7.25}$$

We also know that

$$m \sim N(\mu, \sigma^2/n), \tag{7.26}$$

so that

$$(m - \mu)\sqrt{n}/\sigma \sim N(0, 1). \tag{7.27}$$

Dividing the left-hand side of Equation 7.27 by the square root of the left-hand side of Equation 7.25 over $n - 1$, the number of degrees of freedom, we have

$$\frac{(m - \mu)\sqrt{n}}{\sigma} \times \frac{[\sigma^2(n - 1)]^{1/2}}{[\Sigma(x_i - m)^2]^{1/2}} \sim t_{n-1} \tag{7.28}$$

since, by definition, the ratio of a normal variate to the square root of a χ^2 variate divided by the number of degrees of freedom follows a t distribution (Equation 5.33). But, s_m, the standard deviation of the mean, is

$$s_m = \left[\frac{\Sigma(x_i - m)^2}{n(n - 1)}\right]^{1/2} \tag{7.29}$$

so that

$$(m - \mu)/s_m \sim t_{n-1}. \tag{7.30}$$

The importance of this result is that we do not need to know μ, the true value of the mean, since that will be determined by the hypothesis under test. But σ cancels in Equation 7.28, so we do not need to know its value either.

7.10.2 t test for samples of unequal variance

If the true variances or the number of measurements contributing to the means of two independent samples are not equal, the parameter

$$t = (m_1 - m_2)/s \qquad 7.31$$

with

$$s^2 = s_1^2/n_1 + s_1^2/n_2 \qquad 7.32$$

does not strictly follow a t distribution. However, Satterthwaite (1946) has shown that if we set $v_i = s_i^2/n_i$, the t statistic given by Equation 7.31 is approximately distributed as t_n, where the number of degrees of freedom is

$$n = (v_1 + v_2)^2/(v_1^2/n_1 + v_2^2/n_2). \qquad 7.33$$

Note that when $v_1 = v_2$ and $n_1 = n_2$, the effective number of degrees of freedom is $n_1 + n_2 - 2$, as expected. However, if $v_1 \ll v_2$, $n = n_2 - 1$. In this case m_1 is effectively known exactly and we are comparing m_2 with a fixed constant so that the appropriate number of degrees of freedom is $n_2 - 1$.

7.10.3 Multiple range tests

Suppose that we have a set of n observations x_i and an acceptance range R corresponding to a significance level α for each observation. Then for any one number, x_i, chosen at random

$$P(x_i \text{ is in } R) = 1 - \alpha. \qquad 7.34$$

Using the law of multiplication

$$P(\text{all } x_i \text{ are in } R) = (1 - \alpha)^n. \qquad 7.35$$

Now if the biggest number is contained in R, all of the numbers must be contained in R so that

$$P(x_{max} \text{ is in } R) = (1 - \alpha)^n, \qquad 7.36$$

and

$$P(x_{max} \text{ is **not** in } R) = 1 - (1 - \alpha)^n. \qquad 7.37$$

For small values of α, the right-hand side of this equation is approximately equal to $n\alpha$. Since $P(x_{max}$ is **not** in $R)$ is the significance level α^*, we have

$$\alpha^* = n\alpha \text{ or } \alpha = \alpha^*/n, \qquad 7.38$$

and for n comparisons we should use a significance level of α^*/n to ensure that all comparisons are significant at a significance level of α^*.

With k numbers we have at most $k(k-1)/2$ independent pairs of differences since we can compare the first number with $(k-1)$ others, the second with $(k-2)$ others, since we have already compared it with the first, and so on. But

$$(k-1) + (k-2)\dots 1 = k(k-1)/2, \qquad 7.39$$

and we see that to obtain the acceptance range at a significance level of α for the difference between the largest and the smallest of a set of k readings, we should use the acceptance range for a single comparison at a significance level of α^*/n where $n = k(k-1)/2$.

7.10.4 The ratio of two variances

Consider a set of measurements from a normal distribution with true mean m and true variance σ^2. As in Appendix, section 7.10.1

$$\Sigma (x_i - m)^2/\sigma^2 \sim \chi^2_{n-1}. \qquad 7.40$$

The left-hand side of Equation 7.40 is $(n-1)s^2/\sigma^2$, so that s^2/σ^2 is a χ^2 variate divided by $(n-1)$. We have also seen in section 5.4.3 that the ratio of two χ^2 variates, each divided by their number of degrees of freedom, defines an F variate so that

$$(s_1^2/\sigma_1^2)/(s_2^2/\sigma_2^2) \sim F_{n_1-1, n_2-1} \qquad 7.41$$

where $n_1 - 1$ and $n_2 - 1$ are the number of degrees of freedom associated with s_1 and s_2, respectively. Since our null hypothesis when we compare variances is that $\sigma_1 = \sigma_2$,

$$s_1^2/s_2^2 \sim F_{n_1-1, n_2-1}. \qquad 7.42$$

7.11 EXERCISES

1. In Student's study of the effects of hyoscyamine and hyoscine (Table 7.2), we used t tests to decide if the drugs were efficacious. Would our conclusions have been different if we had used a normal distribution rather than t distributions to determine the critical values for each significance level?

2. Determine what the significance levels in Table 7.10 would have been if we had used the normal distribution instead of the t distribution.

3. The scutum widths of ticks collected from four cotton-tail rabbits were measured and the data are given in Table 7.11 (Sokal and Rohlf, 1981, p. 211). Using Satterthwaite's rule (Appendix, section 7.10.2, Equation 7.33) to determine the appropriate number of degrees of freedom, carry out t tests to decide which, if any, of the mean scutum widths differ significantly. Make the comparisons with and without a correction for the number of comparisons (section 7.4).

4. In Table 7.9 the standard deviation of the weight gained by male rats feeding on

Table 7.11 Number of ticks, mean scutum width in microns and standard deviation of the mean width in microns, taken from each of four rabbits

Rabbit	n	m	s_m
1	8	372.3	6.23
2	10	354.4	2.60
3	13	355.3	3.77
4	6	361.3	2.47

fresh lard is 0.33 g while the weight gained by female rats feeding on fresh lard is 15.96 g. Our analysis assumes that the variances do differ significantly. Carry out an F test to see if this assumption is justified. When deciding on the significance level, remember that you are comparing the largest and the smallest of four numbers. ($F_{2,2}$ at the 0.42% significance level is 237 and at the 0.083% significance level is 1200.)

5. How many patients would you need to use in order to decide if hyoscyamine increases the amount of sleep by at least 30 minutes while keeping the probability of making a Type I error to less than 5% and the probability of making a Type II error to less than 10%?

6. Our analysis showed that a trial of the polio vaccine designed to detect a reduction of at least 50% in the number of paralytic cases required 83 000 children in each group if the probability of making a Type I error is to be less than 1% and the probability of making a Type II error is to be less than 5%. However many children would you need to include in your trial if you wanted to keep the probability of making a Type II error less than 1%?

8

Analysis of variance

...Like every other human endeavor [logic] is just a patchwork quilt whose patches do not meet very well, and which are continually being torn up and restitched.

I. Hacking (1979)

In Chapters 6 and 7 we tested hypotheses about sets of data by comparing means and using the standard deviation as a measure of their precision. For example, the number of children contracting polio was significantly reduced when they were given the Salk vaccine and patients given hyoscine slept for significantly longer than patients given hyoscyamine. Rats given fresh lard gained significantly more weight than rats given rancid lard. This is a sound and reliable way to proceed. However, we have already seen that when we have several treatments each with several levels, such as male and female rats feeding on fresh and rancid lard, the analysis rapidly becomes complicated.

Let us consider an alternative approach. We have an experiment in which we apply several treatments, each at several levels. We know that even if none of the treatments we apply has a significant effect there will still be some variation in our data associated with each level of each treatment due to the random nature of the sampling and measurement process. In this chapter we will show that we can partition the variation of all the data about the grand mean into contributions associated with each treatment factor plus a residual term. By comparing the variation that we attribute to each treatment factor with the residual variation we will be able to decide which factors contribute significantly to the overall variation of the data.

In our experiment on rats, for example, we will consider the freshness of the lard and the sex of the rats in turn and see how much of the observed variation in the weight gain we can explain away using each factor. Instead of asking: 'Do rats gain more weight when the lard is fresh than they do when it is rancid?' we ask instead: 'How much of the variation in weight gain can be explained on the basis of the freshness of the lard?' If very little of the variation can be accounted for using lard freshness, we conclude that lard freshness does not influence the weight gain of the rats significantly. If much of the variation can be accounted for using lard freshness, we conclude that lard freshness does influence the weight gain of the rats significantly and we then return to the kind of analysis we carried out before and ask: 'Do male rats gain significantly more weight than female rats?' Do rats gain more weight when the lard is fresh or when it is rancid? What the **analysis of variance**, or **ANOVA**, does for us is to test each factor and combination of factors (freshness of lard, sex of rats, the interaction between then), enabling us to pick out the significant factors for further consideration. If the ANOVA indicates that certain factors are significant, we then compare the different levels within each treatment factor using the kind of analysis developed in Chapters 6 and 7.

In this book I have not given the computational formulae needed to carry out ANOVAs. The calculations are specific to each experimental design and are given in most of the standard books on statistics. The calculations given in this and the next chapter can be carried out using standard statistical packages and you should do them yourself as you read the book.

8.1 ONE-WAY ANOVA

The central role that ANOVA plays in the analysis of biological data is due to the fact that the sums of the squares of the residuals in suitably designed experiments, calculated without regard to the various treatments or factors involved, can be broken down into a sum of terms, one for each factor involved. To illustrate this we will begin by examining the simplest possible experimental design.

8.1.1 Analysing the sum of squares

Consider a treatment or factor with two levels, T_1 and T_2. We make two measurements at each level, a and c, at level T_1, b and d, at level T_2, as illustrated in Table 8.1. The two levels might be two different cows, in which case a and c might be two estimates of the number of ticks on cow 1; b and d two estimates of the number of ticks on cow 2. Alternatively, the two levels might be two designs of tsetse fly traps in which case a and c might be the number of flies caught on two days in trap 1, b and d the number of flies caught on the same two days in trap 2. Or the two levels might be two kinds of fertilizer and the measurements two estimates of a maize harvest after the application of each kind of fertilizer.

We proceed as follows.

- First we calculate the overall mean and the overall or **total sums of squares**, SS_T, of the deviations about the overall mean for all of the observations, ignoring the fact that we have two treatments.
- We then partition the total sums of squares into a part due to differences **between treatments**, SS_B (arising from differences between cows, traps or fertilizers) and a part due to differences **within treatments**, SS_W, (arising from variability in our sampling of ticks, flies or maize).
- For each sum of squares we calculate a corresponding variance, MS_B and MS_W.
- If the between-treatment variance, MS_B, is significantly **greater** than the within-treatment variance, MS_W, we conclude that the two levels, T_1 and T_2, of the factor or treatment do have a significant effect on the measurements (the cows, traps or fertilizer differ), but if MS_B is **less** than or about the same as MS_W, we

Table 8.1 A model of an experiment in which we apply treatment T at two levels, T_1 and T_2, and make two measurements of the response, a and c, b and d, at each level of each treatment. The table shows the breakdown of the total sum of squares, SS_T, into SS_W, the sum of squares within, and SS_B, the sum of squares between, groups of measurements. MS indicates mean square, $d.f.$ degrees of freedom

T_1	T_2	
a	b	
c	d	
m_1	m_2	SS_B
SS_1	SS_2	SS_T

        ①   ②

- $SS_T = (a^2 + b^2 + c^2 + d^2) - (a + b + c + d)^2/4$
  $d.f. = 4 - 1 = 3$ $\qquad\qquad\qquad MS_T = SS_T/3$

- $SS_1 = (a^2 + c^2) - (a + c)^2/2$
  $d.f. = 2 - 1 = 1$ $\qquad\qquad\qquad MS_1 = SS_1/1$
  $SS_2 = (b^2 + d^2) - (b + d)^2/2$
  $d.f. = 2 - 1 = 1$ $\qquad\qquad\qquad MS_2 = SS_2/1$

                  ①   ③   ④

$SS_W = SS_1 + SS_2 = (a^2 + b^2 + c^2 + d^2) - (a + c)^2/2 - (b + d)^2/2$
$d.f. = (2 - 1) + (2 - 1) = 2$ $\qquad\qquad MS_W = (SS_1 + SS_2)/2$

             ③   ④   ②

- $SS_B = \{m_1^2 + m_2^2 - (m_1 + m_2)^2/2\} \times 2 = (a + c)^2/2 + (b + d)^2/2 - (a + c + b + d)^2/4$
  $d.f. = 2 - 1 = 1$ $\qquad\qquad\qquad MS_B = SS_B/1$

conclude that the two levels, T_1 and T_2, of the two factors or treatments do not have a significant effect on the measurements (the cows, traps or fertilizers do not differ.)

- If the ANOVA indicates that the two levels of the treatment or factors produce significantly different results, we analyse the data further (compare individual cows, traps or fertilizers) and decide what the differences mean.

Table 8.1 gives a step-by-step breakdown of the analysis of variance for a **single classification**, or **one-way**, **two-level** experiment with two repeated measures. The first calculation is for SS_T, the total sum of squares for all of the readings about the mean of all of the readings: this is what we need to account for. To calculate this, remember that the sum of squares of the deviations about the mean can be written as (Equation 4.11)

$$SS = \Sigma x_i^2 - (\Sigma x_i)^2/n. \qquad\qquad 8.1$$

There are four measurements and we have used up 1 degree of freedom in calculating the mean, leaving 3 degrees of freedom. If the treatments have no effect,

MS_T, the total mean square, will be an estimate of the variance due to the intrinsic variations in the readings.

The second calculation is for the sum of squares within each treatment about their means m_1 and m_2: SS_1 for treatment T_1, SS_2 for treatment T_2 and SS_W for both treatments together. Each of these has two measurements and we use up 1 degree of freedom in calculating each mean, leaving 1 degree of freedom for each treatment. SS_W therefore has 2 degrees of freedom and MS_W is our best estimate of the within-group variance.

The third calculation is for SS_B, the sum of the squares between the means m_1 and m_2, of the two groups about their common mean. Since we have two measurements within each group, we multiply by two. This time we have two means and we calculate their overall mean, leaving 1 degree of freedom. MS_B is an estimate of the between-group variance.

The first thing to note is that

$$SS_T = SS_W + SS_B. \qquad 8.2$$

We can see this directly by substituting the expressions given in Table 8.1 into Equation 8.2. The terms labelled ① and ② appear on both sides of the equation and the terms labelled ③ and ④ cancel. In other words, it is possible to partition the **total sum** of squares into a sum of squares **within** groups and a sum of squares **between** groups. (Snedecor and Cochran, 1989, p. 225, give more general proofs of Equation 8.2 for this and for other experimental designs.)

The second thing to note is that the number of degrees of freedom for SS_T, 3, is equal to the number of degrees of freedom for SS_W, 2, plus the number of degrees of freedom for SS_B, 1, so that

$$d.f._T = d.f._W + d.f._B. \qquad 8.3$$

Dividing each sum of squares by the appropriate number of degrees of freedom gives three variances—MS_T, MS_W and MS_B—and if there is no treatment effect, all three are estimates of the variance due to the intrinsic variation in the data. In this ANOVA, the hypothesis that there is no treatment effect reduces to a test of the hypothesis that $MS_W = MS_B$.

To be clear about the calculation of the number of degrees of freedom, note that for a one-way ANOVA with a levels and n measures, we calculate the numbers of degrees of freedom for each sum of squares as indicated in Table 8.2.

Table 8.2 Calculation of the number of degrees of freedom in a one-way ANOVA with a levels each with n measurements. Note that $d.f._T = d.f._W + d.f._B$

SS_T: an points $- 1$ mean $\rightarrow an - 1$ d. f.

SS_W: n points $- 1$ mean $\rightarrow n - 1$ d. f.
Summed over a levels $\rightarrow a(n - 1)$ d. f.

SS_B: a levels $- 1$ mean $\rightarrow a - 1$ d. f.

8.1.2 No treatment effects

The observation that all three variances are estimates of the overall variance of the data, provided there is no treatment effect, is critical. To illustrate this, I generated 18 normally distributed random numbers from $N(0,1)$ and divided them into nine pairs (Table 8.3). We can now analyse this is as a one-way ANOVA with nine levels and two measurements per level.

Table 8.4 gives the ANOVA for the 'without treatment' data in Table 8.3 (columns 2 and 3). The sums of squares between and within groups add up to the total sum of squares and the number of degrees of freedom between and within groups add up to the total number of degrees of freedom. The sums of squares divided by the appropriate numbers of degrees of freedom give the mean squares. Since there is no treatment effect, each gives a reasonable estimate of the true variance, which we know to be 1.

8.1.3 Treatment effects

What we really want to know is what happens when the treatment does have an effect. To illustrate this I took the nine pairs of numbers in Table 8.3 (columns 2 and 3), subtracted 4 from the first pair of numbers, 3 from the second pair and so on,

Table 8.3 Eighteen normally distributed random numbers with mean zero and variance 1 (columns 2 and 3). In columns 4 and 5, 4 has been subtracted from the first pair, 3 from the second pair ..., 4 has been added to the last pair

| Treatment | Without treatment effect | | With treatment effect | |
	1	2	1	2
1	− 1.02	0.23	− 5.02	− 3.77
2	− 0.24	0.01	− 3.24	− 2.99
3	2.18	− 0.22	0.18	− 2.22
4	− 0.47	0.19	− 1.47	− 0.81
5	− 0.74	0.10	− 0.74	0.10
6	0.89	1.98	1.89	2.98
7	− 0.98	− 1.17	1.02	0.83
8	0.23	0.31	3.23	3.31
9	0.42	0.23	4.42	4.23

Table 8.4 ANOVA table for nine pairs of random numbers chosen from $N(0, 1)$

Source of variation	Sums of squares	Degrees of freedom	Mean square	F-ratio	Significance level, P	*
Between	6.6	8	0.82	0.65	0.72	ns
Within	11.4	9	1.27			
Total	18.0	17	1.06			

Table 8.5 ANOVA for the data in columns 4 and 5 in Table 8.3

Source of variation	Sums of squares	Degrees of freedom	Mean square	F-ratio	Significance level, P	*
Between	132.2	8	16.53	13.0	0.00042	***
Within	11.4	9	1.27			
Total	143.6	17	8.45			

adding 4 to the last pair. The data with the 'treatment' effect are also given in Table 8.3 (columns 4 and 5) and the resulting ANOVA is given in Table 8.5. The between groups sum of squares has increased dramatically, the within groups sum of squares has not changed and the residual sum of squares has increased by the same amount as the between groups sum of squares.

The partitioning of the sums of squares is carried out as before. The treatment effect will not change any of the within treatments sums of squares since all of the numbers, as well as the mean, for a given treatment level, will be changed by the same amount and the sum of squares of the deviations about the mean will not change. However, we can show (Snedecor and Cochran, 1989) that, on average, the between groups sum of squares will increase by $(a-1)ns_t^2$ where a is the number of levels of the treatment, n is the number of measurements made at each level and s_t^2 is the population variance calculated for the effect of the different treatment levels on the outcome. Since $(a-1)$ is the number of degrees of freedom associated with the treatment, the between groups mean will increase by ns_t^2. (This is discussed further in section 8.4.6.)

In the example given in Table 8.3 the treatment effect changes the individual levels by $-4, -3 \ldots 2, 3, 4$. The sums of squares for the treatment effect is therefore $16+9+\ldots4+9+16=60$ and since two measurements are made at each level the sum of squares for the treatment effect should increase by about 120. From Tables 8.4 and 8.5 we see that the actual increase is 125.6. The expected increase in the between levels mean square is $120/8 = 15$ and we see that the actual increase is 15.7. Finally, we can estimate the population variance of the levels for a given treatment by dividing the increase in the between levels mean square by two, the number of repeats per level of treatment. This gives 7.9 which compares well with the actual value of $60/8 = 7.5$. The population variance of the treatment levels is referred to as the added variance component per level of that treatment.

8.1.4 Testing hypotheses

It is evident that the mean square estimates of the variance in Table 8.4 are approximately equal, indicating that there is no 'treatment' effect, while in Table 8.5 the between-groups mean square is so much greater than the within-groups mean square that we can be confident that in this case there is a significant treatment effect. Once again, we want to set significance levels to these statements, starting with the null hypothesis that there is no significant treatment effect.

If there is no significant treatment effect, the between-groups mean square, MS_B, and the within-groups mean square, MS_W, are both estimates of the intrinsic variance of the data, σ^2, the first with $a - 1$ degrees of freedom the second with $a(n - 1)$ degrees of freedom (Table 8.2), so that

$$SS_B/\sigma^2 \sim \chi^2_{a-1}, \quad SS_W/\sigma^2 \sim \chi^2_{a(n-1)} \tag{8.4}$$

and if the null hypothesis holds, MS_B/MS_W will follow an F distribution with $a - 1$ and $a(n - 1)$ degrees of freedom. We therefore compare MS_B/MS_W with critical values for the appropriate F distribution. For our simulated experiment with random numbers, we have 8 and 9 degrees of freedom. For the data in Table 8.4, the F ratio (0.65) is well within the 5% acceptance range for $F_{8,9}$ (3.23) and we write *ns* for 'not significant' in the last column; for the data in Table 8.5, the F ratio (13.0) is outside the 0.1% acceptance range for $F_{8,9}$ (10.4) and we give it three stars. Tables 8.4 and 8.5 also give the significance level corresponding to an acceptance range that just includes the observed point. Thus in the case of Table 8.4, an acceptance range from 0 to 0.65, for an F distribution with 8 and 9 degrees of freedom corresponds to a 72% significance level, while in Table 8.7 the acceptance range from 0 to 13.0 corresponds to a significance level of 0.04%.

8.1.5 Means plots

Having carried out the ANOVA and decided that something interesting is going on in our second set of data, we have to decide what this is. Plotting the means for each treatment, as shown in Fig. 8.1, it is clear that the mean increases from one

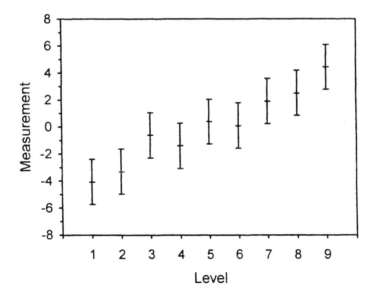

Figure 8.1 Means plot for the data used in Table 8.7. The error bars indicate $\pm s_m$.

treatment to the next. If this were real data, we would want to decide why the mean increases with treatment as it does. When we carry out an ANOVA we start by testing each factor or treatment to see if it has a significant effect on our observations; if it does we then examine the individual levels of that treatment to decide how the levels within that treatment factor differ in their effect. This is more efficient than calculating the means for each pair of points at each level and then making all possible comparisons to decide which, if any, differ.

8.1.6 Soporific drugs

Let us reconsider the data analysed by Student on the effect of hyoscyamine and hyoscine on the amount of sleep gained by ten patients given in Table 7.2. We will begin (as in section 7.2.1) by ignoring the fact that each patient received both drugs. The two drugs are then two levels of treatment with ten repeats at each level. This is a nested design with patients nested in drugs (see section 7.8.1). If we carry out a single-classification ANOVA on the data, we obtain the results shown in Table 8.6. From the ANOVA table we see that there is no significant difference between treatments at the 5% level, which is what we found in section 7.2.1 (Equation 7.3) when we compared the two means using the unpaired t test. Indeed, the value of t was 1.86, which is the square root of 3.5, the value of F given in Table 8.6, showing that in a single-classification experiment the t test provides exactly the same answer as the ANOVA.

Table 8.6 gives both the internal (s_i) and the pooled (s_p) standard deviations of the means. The internal standard deviation is the standard deviation calculated separately for each of the two treatments and the values in Table 8.6 agree with the values calculated in Table 7.2. The pooled standard deviation is the root-mean square value of the two internal standard deviations. Since the within-groups mean square, 3.61, is an estimate of the variance of the response from one patient to the next, we can also calculate the pooled standard deviation as $(3.6/9)^{1/2} = 0.60$. When comparing responses under different levels of the same treatment, we use the pooled standard deviation because this gives us the best estimate of the within-groups variation. (If

Table 8.6 ANOVA table for the two treatments in Student's data of Table 6.6. s_i is the internal standard deviation and s_p is the pooled standard deviation

Source of variation	Sums of squares	Degrees of freedom	Mean square	F-ratio	Significance level, P	*
Between	12.5	1	12.5	3.5	0.079	ns
Within	64.9	18	3.61			
Total	77.4	19				

	Treatment	Mean	s_i	s_p	95% C.L.
	Hyoscyamine	0.75	0.57	0.60	−0.51−2.01
	Hyoscine	2.33	0.63	0.60	1.06−3.59

the internal standard deviations differ significantly, we have to be more careful and use Satterthwaite's rule for the t test as recommended in section 7.2.1.)

To calculate 95% confidence limits for the amount of sleep gained, we multiply the standard deviations by the two-tailed value of t_{18} at the 5% significance level, 2.10, so that the amount of sleep induced by hyoscyamine, for example, is $0.75 \pm 0.6 \times 2.10$, that is from -0.51 to 2.01. In Table 8.6 the 95% confidence intervals for the two means overlap, reinforcing our conclusion that they do not differ significantly. We also note that the confidence interval for the first mean includes 0, while that for the second does not, and this is again in agreement with our previous observation that hyoscyamine does not significantly increase the amount of sleep while hyoscine does.

8.2 UNEQUAL SAMPLE SIZES

When we have a non-way ANOVA with only a few levels it may be sufficient to calculate means and standard deviations and then compare the means using a suitable multiple range test if there is any doubt as to the significance of the various differences. The real power of ANOVA will be apparent only when we analyse experiments with more than one treatment or factor. However, it is important to have a firm grasp of the basic ideas, so we will consider another example of a one-way ANOVA but with unequal sample sizes.

An experiment was performed in which tick larvae of the species *Haemaphysalis leporispalustris* were collected from four cotton-tail rabbits (Sokal and Rohlf, 1987,

Table 8.7 Scutum widths, in microns, of tick larvae taken from four cottontail rabbits

	Rabbit			
	1	2	3	4
	380	350	354	376
	376	356	360	344
	360	358	362	342
	368	376	352	372
	372	338	366	374
	366	342	372	360
	374	366	362	
	382	350	344	
		344	342	
		364	358	
			351	
			348	
			348	
m	372.3	354.4	355.3	361.3
s_m	6.23	2.60	3.77	2.47

Table 8.8 ANOVA table for the data of Table 8.8. 'Among' refers to differences among rabbits and 'within' to differences among ticks on each rabbit

Source of variation	Sums of squares	Degrees of freedom	Mean square	F-ratio	Significance level, P	*
Among	1808	3	603	5.26	0.0045	**
Within	3778	33	115			
Total	5586	36	102			

p. 217). The width of the scutum, the dorsal shield, of each larva was measured in microns and the data are given in Table 8.7. The mean scutum width is about the same for rabbits 2 and 3, while for rabbit 4 it is a little greater and for rabbit 1 it is greater still. The difference between the biggest and the smallest mean is 17.9, but this is less than three times the standard deviation of the mean for larvae from rabbit 4, and we will need to be careful in our analysis. The analysis of variance is shown in Table 8.8 and Sokal and Rohlf (1987, p. 168) perform the calulation step by step.

The number of degrees of freedom among rabbits is $4 - 1 = 3$. Since rabbit 1 has eight ticks, the number of degrees of freedom 'within' rabbit 1 is $8 - 1$ and taking all the rabbits together the number of degrees of freedom 'within' rabbits is $(8 - 1) + (10 - 1) + (13 - 1) + (6 - 1) = 33$. The total number of degrees of freedom is $37 - 1 = 36$. Both the sums of squares and the number of degrees of freedom between and within groups shown in Table 8.8 add up to the corresponding totals. The F ratio is significant at the 1% level and has two stars.

So far, all that we know is that the mean scutum width of the ticks differ among the four rabbits. Now we need to consider individual rabbits. If we plot the mean scutum widths for the ticks on each rabbit together with 95% confidence intervals,

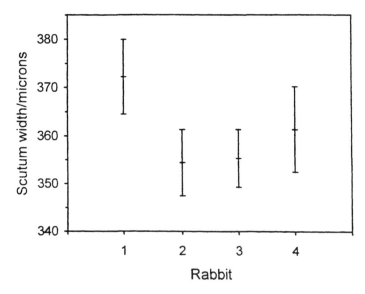

Figure 8.2 Mean scutum width/microns for the larvae from each of four rabbits.

Table 8.9 Comparison of the mean scutum widths from the data in Table 8.7. The columns of stars and letters indicate which means differ significantly using an LSD test (section 7.4) at the 95% significance level

Rabbit	Number of larvae	⟨scutum width⟩ (microns)	Groups	Groups
2	10	354.4	*	a
3	13	355.3	*	a
4	6	361.3	* *	ab
1	8	372.3	*	b

as shown in Fig. 8.2, we see that the confidence limits for the scutum width of larvae from rabbit 1 does not overlap the confidence limits for rabbits 2 and 3, but does overlap the confidence limits for rabbit 4.

In addition to plotting the means such data are often presented as a means table as in Table 8.9. The data are presented in such a way that rabbits sharing a star in the same column do not differ significantly in the scutum widths of the tick larvae taken from them. The first column of stars shows that the mean scutum widths of larvae on rabbits 2, 3 and 4 do not differ, the second shows that the mean scutum widths of larvae on rabbits 4 and 1 do not differ, but since rabbit 1 does not share stars with rabbits 2 or 3, the mean scutum width of larvae on rabbit 1 does differ significantly from those on rabbits 2 and 3. An alternative way of presenting this data is shown in column 5, in which rabbits sharing the same letter do not differ significantly in the mean scutum width of the tick larvae taken from them.

Another useful piece of information we can extract from the ANOVA table (after we have used an F test to check that there is a significant effect) is the contribution to the overall variance arising from a particular treatment. We have already seen (section 8.1.3) that the mean square within groups, MS_W, gives us an estimate of σ^2, the intrinsic variability of the data, while the mean square between groups, MS_B, gives us an estimate of $\sigma^2 + n\sigma_T^2$, where σ_T^2 is the average value per level of the added variance component due to the treatment T and n is the number of levels per treatment, in this case the mean number of ticks per rabbit, i.e. 9.25. Therefore

$$(MS_B - MS_W)/n \approx (\sigma^2 + n\sigma_T^2 - \sigma^2)/n = s_T^2 \qquad 8.5$$

gives us an estimate of this added variance per level. (Sokal and Rohlf, 1981, show how to calculate confidence limits for the added variance.) From the data given in Table 8.8, the average value of the added variance component is 52.8 microns^2 per rabbit and the best estimate of the standard deviation of the variability among rabbits is 7.3 microns.

8.3 MODEL I AND MODEL II ANOVA

The experiment on rabbits discussed in the previous section could have been conducted for either of two reasons. We could have been studying levels of immunity of rabbits to ticks in order to identify the factors that confer such immunity with a view to developing a vaccine against ticks. Rabbits on which ticks grow less well may have some immunity. In this case we would be interested in the individual rabbits and we would refer to the rabbits as a **fixed** factor. When we are interested in individual levels of various treatments or factors, we talk about a **Model I ANOVA**, and we choose particular treatment or factor levels (in this case rabbits) with a view to deciding if the different levels produce significantly different effects. After calculating the ANOVA table, we then go on to test for differences between the means or between various combinations of the means.

We could, however, be interested in the genetic variability within various rabbit populations of their response to ticks so that the variability itself is of interest. When we are interested in the variability across the levels of any one treatment or factor, we talk about a **Model II ANOVA**, and we will generally choose the levels (in this case rabbits) at random, often going to great lengths to ensure that the chosen rabbits are representative of the population at large. We then refer to the rabbits as a **random** factor. In a Model II ANOVA we tacitly assume that the effect of the variability is not only random but is normally distributed. After calculating the ANOVA table, we then go on to assess the added component of the variance arising from the treatment effect. The important point to note is that for a given set of data the precise analysis that we carry out depends on the question that we ask.

We can also have **mixed models** in which some factors are fixed and correspond to Model I and others are random and correspond to Model II. In Student's study of the effects of hyoscyamine and hyoscine on sleep, we hope to determine the effects of these two specific drugs so that the drugs are a fixed factor. Since we are not interested in particular patients but in these patients as representative of the population as a whole, the different patients used in the trial are a random factor and the variance within groups of patients gives us an estimate of the variance in the population as a whole.

8.3.1 Specifying the model

It is often useful to think explicitly about the model that underlies our analysis. Suppose that the response variable is y_{ij}, so that in Student's data on soporific drugs y indicates the number of hours of sleep gained, i indicates each of the two drugs and j indicates each of the individual patients. Then our model is

$$y_{ij} = m + d_i + \varepsilon_{ij} \qquad\qquad 8.6$$

where m is the mean number of hours of sleep gained by all the patients under both treatments, $d_i (i = 1, 2)$ is the change from the mean under each drug and the residuals ε_{ij} give the residual amount of sleep gained by patient $j (j = 1$ to $10)$ not accounted for by m and the drugs d_i. In this case the grand mean tells us that $m = 0.54$ hours and the difference between the means for each drug and the grand mean tell us that $d_1 = 0.21$ hours and $d_2 = -0.21$ hours.

8.4 TWO-WAY ANOVA

The simplest way to extend the one-way ANOVA described in the previous sections is to consider two treatment factors. We have already considered a problem of this kind in the experiment on the diet of rats (section 7.7) where we examined the extent to which the freshness of the lard and the sex of the rats affected the weight gained by the rats.

8.4.1 Factorial designs

Suppose that we have two treatments or factors, A and B (in our experiments on rats these would be lard and sex), each with two levels (fresh/rancid lard, male/female rats). Ideally we would like to make several measurements with every possible combination of A and B: $A_1 B_1$, $A_2 B_1$, $A_1 B_2$ and $A_2 B_2$. When we make measurements on every combination of the relevant factors, we call it a **factorial** experiment. Note that factor A is crossed with factor B since each level of B occurs with each level of A and vice versa.

One way to approach such an experiment would be to treat each combination of factors as though it were a separate factor. We might then use C to represent these combined factors, so that

$$C_1 \equiv A_1 B_1 \quad C_2 \equiv A_2 B_1 \quad C_3 \equiv A_1 B_2 \quad C_4 \equiv A_2 B_2. \qquad 8.7$$

We can then treat this experiment as a one-way ANOVA with C as the single treatment or factor with four levels. In this case, C_1 would be fresh lard, male rats; C_2 would be rancid lard, male rats; C_3 would be fresh lard, female rats; and C_4 would be rancid lard, female rats. Our model would then be

$$y_{ij} = m + c_i + \varepsilon_{ij}. \qquad 8.8$$

Table 8.10 gives the data of Table 7.8 laid out as a one-way ANOVA. The analysis proceeds as before and the resulting ANOVA table is shown in Table 8.11 which shows that the combined treatment factor C, corresponding to sex and lard, is highly significant.

We could now proceed as we did when we were using t tests and examine the effect of sex allowing for the freshness of the lard, and so on. However, it is more

Table 8.10 The weight (in grams) gained by male rats fed on fresh lard C_1 and on rancid lard C_2, and by female rats fed on fresh lard C_3 and on rancid lard C_4

	Treatment		
C_1	C_2	C_3	C_4
171	108	153	85
172	89	109	64
172	69	160	82

Table 8.11 One-way ANOVA for the data of Table 8.13 for the weight gained by male and female rats fed on fresh and on rancid lard with each combination of sex and freshness treated as a separate factor

Source of variation	Sums of squares	Degrees of freedom	Mean square	F-ratio	Significance level, P	*
Among	17 779	3	5926	18.6	0.0006	***
Within	2 548	8	319			
Total	20 327	11				

Table 8.12 The data of Table 8.10 laid out for a two-way ANOVA with three repeated measures

	A_1 (fresh lard)	A_2 (rancid lard)
B_1 (male rats)	171	108
	172	89
	172	69
B_2 (female rats)	153	85
	109	64
	160	82

convenient to build these comparisons directly into our analysis and this is what the two-way (or multi-way) ANOVA does for us. Table 8.12 gives us the data of Table 8.10, but this time laid out as for a two-way ANOVA.

Our model is now

$$y_{ijk} = \mu + a_i + b_j + (ab)_{ij} + \varepsilon_{ijk},$$ 8.9

where a_i indicates the change from the mean value when the lard is fresh or rancid, b_j indicates the change from the mean when the rats are males or females and $(ab)_{ij}$ indicates the effect of the interaction so that $(ab)_{11}$, for example, gives the effect of fresh lard for male rats after allowing for the overall mean, the average effect of sex and the average effect of lard freshness.

To make the details of Model 8.9 explicit we can calculate each term separately. Table 8.13 gives various mean values calculated from the data in Table 8.12. The grand mean is 119.5 g so that in Equation 8.9

$$m = 119.5 \text{ g}.$$ 8.10

The average increase in weight for male rats is 130.2 g so that

$$a_1 = 130.2 - 119.5 = 10.7 \text{ g}$$ 8.11

and in the same way

$$a_2 = 108.8 - 119.5 = -10.7 \text{ g}.$$ 8.12

Table 8.13 Averages calculated from the numbers in Table 8.12. The four numbers in the corners are the means of the four sets of three numbers in Table 8.12. The numbers in italics are the means of the two adjacent numbers (vertically or horizontally) and the number in bold is the grand mean of all the data

	Fresh	⟨Freshness⟩	Rancid
Male	171.7	*130.2*	88.7
⟨Sex⟩	*156.0*	**119.5**	*82.8*
Female	140.7	*108.8*	77.0

The average increase in weight when the lard is fresh is 156.0 g so that

$$b_1 = 156.0 - 119.5 = 36.7\,\text{g} \qquad\qquad 8.13$$

and similarly

$$b_2 = 82.8 - 119.5 = -36.7\,\text{g}. \qquad\qquad 8.14$$

That the two coefficients for the effect of sex and the two coefficients for the effect of lard freshness differ only in signs, reminds us that in each case we have only a single independent number and hence only 1 degree of freedom for each factor.

To calculate the interaction term for male rats eating fresh lard we take the average weight gain for all the rats (119.5 g), add the additional weight gain for male rats (10.7 g), and then add the additional weight gain for fresh lard (36.7 g), to give 166.9 g. Then since the average weight gain for male rats eating fresh lard is 171.7 g, the interaction term is

$$(ab)_{11} = 171.7 - 166.9 = 4.8\,\text{g}. \qquad\qquad 8.15$$

We can calculate the other three interaction terms in the same way:

$$(ab)_{12} = -4.8\,\text{g} \quad (ab)_{21} = 4.8\,\text{g} \quad (ab)_{22} = -4.8\,\text{g}. \qquad\qquad 8.16$$

That the four numbers differ only in their signs shows that we have 1 degree of freedom for the interaction.

The ANOVA for the data of Table 8.12 is given in Table 8.14. Comparing Table 8.14 with Table 8.11, we see that the total and residual sums of squares are the same in both cases and that the sums of squares for the three 'effects'—sex, lard and sex × lard—add up to the sums of squares for the single treatment, 'sex-lard'. In other words, the two-way ANOVA partitions the total sums of squares for the treatments into contributions from the individual treatments and their interactions. In Table 8.14 the effect of sex is not significant at the 5% level, as we found in Table 7.10 using the *t* test. The effect of lard freshness, on the other hand, is highly significant and has three stars, as in Table 7.10. Finally, the effect of the interaction sex × lard is again not significant in agreement with the *t* test. In fact, the two approaches are **entirely** equivalent when we have only two levels for each treatment, so that we are comparing means in pairs. (In Table 7.10 the *t* parameters are 2.07 for sex, 7.12 for lard freshness and 0.94 for the interaction term. If we square these

Table 8.14 Two-way ANOVA for the data of Table 8.12 for the weight gained by male and female rats when fed on fresh and rancid lard

Source of variation	Sums of squares	Degrees of freedom	Mean square	F-ratio	Significance level, P	*
Sex	1 365	1	1365	4.29	0.072	ns
Lard	16 133	1	16 133	50.7	0.0001	***
Sex × Lard	280	1	280	0.88	0.39	ns
Residual	2 548	8	318			
Total	20 326	11				

numbers we have 4.28, 50.7 and 0.88, which are identical to the corresponding F statistics given in Table 8.14.) When we have more than two levels per treatment, the ANOVA assumes that none of them has any effect; if we find a treatment or factor that does have a significant effect, we then proceed to examine its various levels in pairs to discover reasons for the effect. If we were to start from the *t* test, we would have to build up all possible pairs of effects and would rapidly become bogged down in complexity.

In Table 8.14 the sum of squares for sex, lard and sex × lard add up to the sum of squares for the combined treatment sex-lard (Table 8.12), as we would expect. For sex we have two levels, males and females, and we use up 1 degree of freedom when we average over sex leaving us with $2 - 1 = 1$ degree of freedom for sex. In the same way we see that we are left with $2 - 1$ degree of freedom for lard. For the interaction, we have $(2 - 1) = 1$ degree of freedom for sex combined with $(2 - 1) = 1$ degree of freedom for lard, giving us 1×1 degree of freedom for the interaction. For the residual term we have three repeats in each category, giving $(3 - 1) = 2$ degrees of freedom for each one. We repeat this $2 \times 2 = 4$ times for all combinations of sex and lard, giving $2 \times 4 = 8$ degrees of freedom for the residual term. These four numbers 1, 1, 1 and 8 add up to 11 and this is the same as the number of degrees of freedom for the total sum of squares, which is $12 - 1 = 11$.

We can generalize the argument of the preceding paragraph as follows. Suppose that in our experiment treatment A has *a* levels and treatment B has *b* levels. The relevant numbers of degrees of freedom in a two-way factorial ANOVA would then be those given in Table 8.15. For A we have *a* levels and after calculating the deviations from the mean we are left with $(a - 1)$ degrees of freedom. For B we have *b* levels and after calculating the deviations from the mean we are left with $(b - 1)$ degrees of freedom. For $A \times B$ we have $a \times b$ levels, but only $(a - 1)$ and $(b - 1)$ independent levels, giving us $(a - 1) \times (b - 1)$ degrees of freedom. Within each group we have *n* levels and calculate 1 mean, giving us $n - 1$ degrees of freedom, but we repeat this *ab* times, giving us $ab(n - 1)$ degrees of freedom for the residual sum of squares. For the total we take the total number of points *abn* and subtract 1, giving us $abn - 1$ degrees of freedom.

Table 8.15 Decomposition of the number of degrees of freedom in a two-way factorial ANOVA

Treatment	No. of levels	d.f
A	a	$a - 1$
B	b	$b - 1$
$A \times B$	ab	$(a - 1)(b - 1)$
Residual	n	$ab(n - 1)$
Total	abn	$abn - 1$

8.4.2 Multiple comparisons

Making several comparisons brings us back to the discussion of multiple comparisons in section 7.4. If we had decided in advance that we wanted to examine only the effect of the lard, then comparing the F ratios in Table 8.14 with the appropriate critical values is correct. However, if we want to test all three F ratios to see if any of them are significant at a significance level of α, we should use a significance level of $\alpha/3$. The effects of sex and lard are, of course, still not significant. To decide if the effect of lard is significant at the 0.1% level, we need to use a significance level of 0.03% and in this case we see that it still has three stars since the significance level corresponding to the observed F ratio is 0.01%.

8.4.3 Interactions

The roles of the two levels sex and lard in our ANOVA are clear: they tell us if the amount of lard consumed depends on the sex of the rats (after allowing for freshness) or the freshness of the lard (after allowing for sex). We have already discussed the interaction term in Chapter 7 (Table 7.10), where we calculated

$$\langle F - R | m \rangle - \langle F - R | f \rangle, \tag{8.17}$$

that is the difference between the effect of lard freshness on the consumption by male rats and the effect of lard freshness on the consumption by female rats. We also calculated

$$\langle m - f | F \rangle - \langle m - f | R \rangle, \tag{8.18}$$

which is the difference between the effect of the sex of the rats on the consumption of fresh lard and the effect of the sex of the rats on the consumption of rancid lard. We found that the interaction term was the same whichever way we calculated it.

To understand the interaction clearly, consider the hypothetical results given schematically in Fig. 8.3. In Fig. 8.3a the effect of changing A from level A_1 to A_2 is the same for both levels of B, and the effect of changing B from level B_1 to B_2 is the same for both levels of A. The effect of changing A is therefore independent of the level of B, and vice versa, and there is no interaction. In Fig. 8.3b, on the other hand, changing A from level A_1 to A_2 increases the response if we are at level B_1 and decreases the

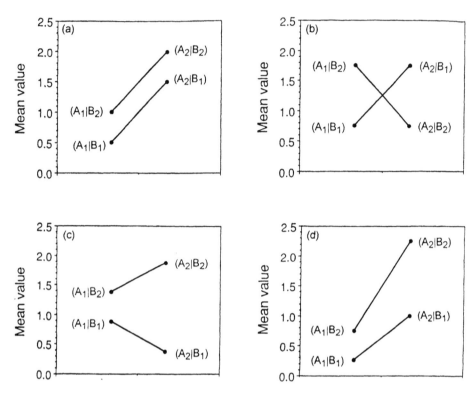

Figure 8.3 Schematic diagram of the means in a two-factor ANOVA. $(A_1|B_1)$, for example, gives the mean value under level 1 of A and level 1 of B. In (a) there is a significant effect for A and B but no interaction $A \times B$. In (b) neither treatment is significant but there is a significant interaction $A \times B$. In (c) treatment B and the interaction are significant but treatment A is not. In (d) both treatments and the interaction are significant.

response if we are at B_2, so that the effect of a change in A depends on the level of B and there is a significant interaction. Similarly, changing B from level B_1 to B_2 increases the response if we are at level A_1 and decreases the response if we are at A_2, so that the interaction is significant. Note, however, that the effect of changing A averaged over the two levels of B is zero and the effect of B averaged over the two levels of A is zero so that neither factor is significant when averaged over the levels of the other factor. Fig. 8.3c shows a situation in which B and the interaction are significant while A is not, while Fig. 8.3d shows a situation in which both treatments and their interaction are significant.

Figure 8.4 is a plot of the mean weight gained by the male and female rats eating fresh and rancid lard and we can see that there is no significant interaction because the two lines are almost parallel.

8.4.4 Soporific drugs

In section 8.16 we analysed Student's data on the effects of two soporific drug on 20 patients treating the patients as a random factor nested in the fixed drugs factor.

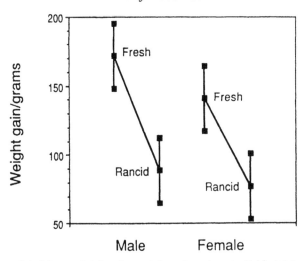

Figure 8.4 Means plot for the weight gains given in Table 8.14.

In fact, only ten patients were used, with each patient being given both drugs and also no drug. Since each patient was given both drugs, we can treat the patients as one factor and the drugs as a second in a factorial design. The patients are still a random factor but they are now **crossed** with the drugs not **nested** under drugs. Since we are not interested in the response of individual patients who have been randomly selected from the overall population, the patients are a Model II factor. We are, however, interested in the effect of each drug separately, and so the drugs are a Model I factor. The model is therefore a mixed-model, two-way ANOVA without replication. The ten patients account for 9 degrees of freedom and the two treatments account for 1 degree of freedom. If we tried to calculate the interaction term, this would have 9 × 1 degrees of freedom, giving us a total of 19, but since we have only 19 degrees of freedom at our disposal, we would have none left for the error term. In other words, we can carry out an ANOVA on a two-factor model without replication, but only if we assume that the interaction term is zero so that we can ignore it.

Carrying out a factorial ANOVA on Student's data with the two factors crossed, we obtain the results shown in Table 8.16. By treating patients as a separate factor, the analysis is equivalent to a paired t test. In Table 7.2 we obtained a value of $t = 4.06$, which is equal to the square root of 16.5, the value of F in Table 8.16, so that the two-way ANOVA without replication is equivalent to the paired t test. By using the fact that each treatment is applied to the same patients twice, we are able to show that the effects of the two drugs are significantly different, whereas in our earlier ANOVA (Table 8.6), in which we did not use this knowledge, the two tests were not significantly different. Of course, if we had used 20 patients and had given each of them one treatment only, the analysis in Table 8.6 would have been the best that we could do. For the two-way ANOVA our model is

$$y_{ij} = m + d_i + p_j + \varepsilon_{ij},$$ 8.19

Table 8.16 ANOVA table for the two drug treatments in Table 7.2. s_i is the internal standard deviation and s_p the pooled standard deviation

Source of variation	Sums of squares	Degrees of freedom	Mean square	F-ratio	Significance level, P	*
Drugs	12.5	1	12.5	16.5	0.0028	**
Patients	58.1	9	6.46	8.53	0.0019	**
Residual	6.8	9	0.756			
Total	77.4	19				

	Drug	Mean	s_i	s_p	95% C.L.
	Hyoscyamine	0.75	0.57	0.27	−0.13–1.37
	Hyoscine	2.33	0.63	0.27	1.71–2.95

where m is the grand mean, d_i is the change from the mean due to drug i, p_j is the change from the mean due to patient j and ε_{ij} is the residual for patient j on drug i.

It is worth comparing Table 8.6, in which we simply treated the patients as replicates, with Table 8.16, in which we treat the patients as a second factor. The 'treatment' sum of squares in Table 8.16 is identical to the 'between-groups' sum of squares in Table 8.6. However, the 'within-group' sum of squares in Table 8.6 is now partitioned into a 'patient' sum of squares and a 'residual' sum of squares. Because we are able to separate these out and compare the treatment effect with the residuals after allowing for the effect of the patients, the denominator in our F test is reduced and the test of our null hypothesis is made more sensitive. The means in Table 8.16 are exactly the same as in Table 8.6. However, the pooled standard deviation in Table 8.16 (0.27) is less than the value of the pooled standard deviation in Table 8.6 (0.60) as it is now calculated after allowing for the variation among patients. The reduction in the pooled standard deviation increases the sensitivity of the comparison.

8.4.5 Latin squares

We would always like to make measurements for all possible combinations of the levels of each factor and with several repeats of each measurement to that we can assess the effect of each factor as well as all the interactions among the factors. Unfortunately this is not always possible. We have already seen in our study of the polio vaccine that we cannot give the vaccine and the placebo to each child and we will return to this again in the next section. Consider, however, the following problem. Tsetse flies (*Glossina* spp.) are the vectors of trypanosome diseases called *nagana* in cattle and sleeping sickness in people. Trypanosomiasis is widely regarded as one of the most debilitating diseases in Africa and where tsetse flies are present livestock are largely absent. This makes it impossible to keep livestock in much of the potentially most productive areas of Africa. The lack of livestock results in a lack of protein, milk, tractive power, fertilizer and leather goods. Recent work, however, has shown that testse fly populations can be controlled effectively using traps

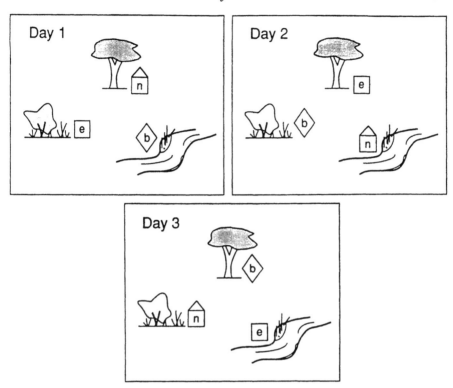

Figure 8.5 Three traps, b — biconical, e — epsilon and n — Nguruman, are rotated around three different sites on three different days in a Latin square experiment.

(Dransfield *et al.*, 1991). The development of traps that are effective against tsetse flies has involved testing many different trap shapes, colours and attractive odours. How can we best compare different traps in field conditions? We know that the number of flies that we catch in a trap varies significantly from day to day, for reasons that are still not understood. We also know that the number of flies that we catch in a trap depends on the site in which we put the trap, probably because of variations in the visibility and degree of shading at different sites. Now we can put trap 1 into site 1 on day 1 but we cannot then put another trap into the same site on the same day, and we are faced with the problem that we cannot carry out measurements on all possible combinations of treatments. In this situation a very useful design is the Latin square, illustrated in Fig. 8.5. We see that we can rotate the traps between sites on successive days so that each trap occupies each site over the 3 days. (Note that the square refers to the design and not to the actual layout although the layout could also be a square.)

Dransfield and Brightwell (pers. comm.) carried out an experiment to compare the effectiveness of three traps for female *Glossina pallidipes*. The first and oldest trap is called the biconical because it is made out of two cloth cones joined at their bases (Challier *et al.*, 1977). A more recently developed trap is the Nguruman trap (Brightwell

Table 8.17 Numbers of tsetse flies caught in biconical, *b*, epsilon, *e*, and Nguruman, *n*, traps. Three Latin squares were used. Within each square there were three sites and measurements were taken over 3 days so that all three traps are placed once in each site and on each day all three traps are found in one of the three sites for that square

	Day		
Site	1	2	3
1	*n* 393	*e* 230	*b* 158
2	*b* 58	*n* 146	*e* 392
3	*e* 159	*b* 60	*n* 392
4	*e* 97	*b* 22	*n* 252
5	*n* 443	*e* 119	*b* 81
6	*b* 33	*n* 166	*e* 162
7	*b* 12	*n* 103	*e* 38
8	*e* 55	*b* 30	*n* 216
9	*n* 89	*e* 132	*b* 44

et al., 1991) and this consists of a triangular prism made of cloth with a netting cone on the top. The third trap is the epsilon trap which is similar to the Nguruman trap with the cone recessed into the prism. All three traps were baited with acetone, cow urine and octenol, which are known to be attractive to tsetse flies. The results of the experiment are recorded in Table 8.17.

The Latin square is an incomplete factorial design with all factors crossed. Before we carry out the analysis of variance we need to consider the model that we wish to apply to our data. We assume that the effects of sites, traps and days are multiplicative so that if site 2 is twice as good as site 1, if there are twice as many flies present on day 2 as there are on day 1, and if trap 2 is twice as good as trap 1, then trap 2 on day 2 in site 2 should catch $2 \times 2 \times 2 = 8$ times as many flies as trap 1 in site 1 on day 1. However, analysis of variance assumes that the effects are additive. We therefore take the logarithm of the trap catches since the logarithm of a product is the sum of the logarithms and multiplicative effects become additive after transforming the catches to logarithms. Our model is therefore

$$log(\text{catch}) = m + s_i + t_j + d_k + \varepsilon_{ijk}, \qquad 8.20$$

where s_i refers to the three sites, t_j to the three traps and d_k to the three days.

We can now analyse each square separately and the ANOVA table for the first square (sites 1, 2 and 3 in Table 8.17) is given in Table 8.18 from which we see that the epsilon trap catches more flies than the biconical trap and the Nguruman trap catches more flies that the epsilon trap. However, the only significant difference is between the biconical trap and the Nguruman trap. If we repeat the exercise for the other squares, then in square 2 the biconical trap differs significantly from both the Nguruman and the epsilon while in square 3 none of the traps differs significantly from any other. The reason for this is that in square 3 fewer flies were caught than

Table 8.18 Analysis of variance for the first Latin square given in Table 8.17. The catches were transformed by calculating the logarithm of the catch (base 10) before carrying out the analysis of variance. 'Lower' and 'upper' give 95% confidence limits for the transformed mean catches. The means are compared using Tukey's HSD at the 95% level

Source of variation	Sums of squares	Degrees of freedom	Mean square	F-ratio	Significance level, P	*
Site	0.2091	2	0.1045	21.4	0.040	*
Trap	0.4364	2	0.2182	44.7	0.022	*
Day	0.0594	2	0.0297	6.08	0.141	ns
Residual	0.0098	2	0.0049			
Total	0.7147	8				

Pooled standard error = 0.0403

Trap	Lower	Mean	Upper	
Biconical	1.740	1.913	2.087	a
Epsilon	2.050	2.223	2.397	ab
Nguruman	2.277	2.451	2.624	b

in the other two squares. Note that we use Tukey's HSD test since we are interested in comparing all three traps.

The problem with a single 3 × 3 Latin square is that we have 2 degrees of freedom for each of our three factors but only 8 degrees of freedom altogether so that we are left with only 2 degrees of freedom for the error term. We would like to combine the three squares into one experiment. We have to be careful as to how we do this. In this particular experiment the three squares were run on the same three days so that we can treat the three squares as one Latin rectangle with the three traps spread over 3 days and nine sites.

Table 8.19 gives the analysis of variance for the data of Table 8.17 treated as a Latin rectangle. By combining the three squares, we have 14 degrees of freedom for the residual. The site effect and the trap effect are both significant at the 0.1% level and get three stars while the day effect is not quite significant at the 5% level. Furthermore, we now see that all three traps differ significantly, with the epsilon trap significantly better than the biconical trap and the Nguruman trap significantly better than the epsilon. It is worth noting that the pooled standard error in the Latin rectangle, 0.0491, is slightly greater than the pooled standard error for square 1. However, to calculate 95% confidence limits for the individual squares where we have 2 degrees of freedom for the error term, we multiply by $t_2(0.975) = 4.30$, while for the Latin rectangle where we have 14 degrees of freedom for the error term we multiply by $t_{14}(0.975) = 2.15$, and this makes the test on the combined data significantly more sensitive than the tests for each of the individual squares. If the number of degrees of freedom remaining for the error term is very small, all tests are going to be rather insensitive.

Since the biconical trap is well established, it is often used as a standard against

Table 8.19 Analysis of variance for the data in Table 8.17 treated as a Latin rectangle. The catches were transformed by calculating the logarithm (base 10) of the catch before carrying out the analysis of variance. 'Lower' and 'upper' give 95% confidence limits for the transformed mean catches. The means are compared using Turkey's HSD at the 95% level

Source of variation	Sums of squares	Degrees of freedom	Mean square	F ratio	Significance level, P	*
Site	1.394	8	0.174	7.76	0.0005	***
Trap	2.165	2	1.082	48.2	0.0000	***
Day	0.158	2	0.079	3.52	0.058	ns
Residual	0.314	14	0.0224			
Total	4.031	26				

Pooled standard error = 0.0499

Trap	Lower	Mean	Upper	
Biconical	1.527	1.634	1.742	a
Epsilon	1.935	2.042	2.150	b
Nguruman	2.217	2.324	2.431	c

which to compare other traps and we can calculate an index of increase by detransforming the means and then calculating the ratio of the catch in each trap to the catch in the biconical trap and this gives us the result in Table 8.20.

To determine error limits for the indices of increase, we calculate errors for the differences between the transformed mean catches as the square root of the sums of the squares of the errors for each pair of traps. We then add the error term to the difference between the transformed mean and detransform the result to obtain the upper limit and subtract the error term and proceed in the same way to obtain the lower limit. Although the errors in the differences of the transformed means are symmetrical, the errors in the detransformed means are not, as we can see in Table 8.20. We see then that the epsilon trap catches, on average, 2.5 times as many flies as the biconical with 95% confidence limits from 1.5 to 4.2, while the Nguruman trap catches, on average, 4.8 times as many flies as the biconical with 95% confidence limits ranging from 2.9 to 7.9.

Table 8.20 Indices of increase for the epsilon and Nguruman traps over the biconical trap with 95% confidence limits

	Lower	Mean	Upper
Biconical	●	1.00	●
Epsilon	1.51	2.51	4.16
Nguruman	2.88	4.78	7.91

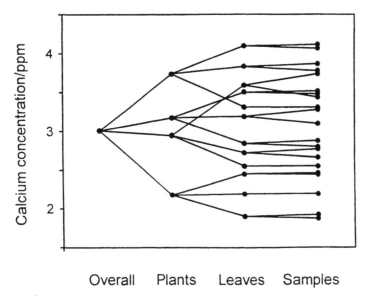

Figure 8.6 Schematic illustration of the dependence of the calcium concentration on the plants, the leaves and the individual calcium determinations. The data are in Table 8.21.

8.4.6 Nested designs

We have seen that in a factorial design we make measurements for all possible combinations of the levels of each factor. A Latin square design is an incomplete factorial design that we use when we cannot set up all possible combinations of the levels of each factor. Another important class of designs arises when factors are nested within one another. In the analysis of Student's data on soporific drugs, we noted that if we had 20 different patients we would have performed either an unpaired t test or a single-factor ANOVA; the patients would have been nested within the drugs. In the analysis of scutum widths of tick larvae on rabbits the ticks were nested within rabbits and we were able to estimate the added variance component due to rabbits.

Consider the following example of a nested design, discussed in detail by Snedecor and Cochran (1989, p. 247), in which the calcium concentration in the leaves of turnip plants was measured as a percentage of the dry weight of the leaves. Four turnip plants were chosen; from each plant three leaves were chosen; from each leaf two samples were taken and used to measure the concentration of calcium. Clearly, we might expect the calcium concentration to vary between plants, the concentration within each plant to vary between leaves and the concentration within each leaf to vary between the two measurements of the calcium concentration. This hierarchical structure is illustrated in Fig. 8.6 which shows the data given in Table 8.21.

The various factors are nested one within the other, because the same leaf cannot be on more than one plant and each sample cannot be from more than one leaf. The model for this design is written as follows

$$y_{ijk} = m + p_i + l_{ij} + s_{ijk},$$ 8.21

Table 8.21 Four plants were chosen (column1). From each of these three leaves were chosen (column 2). From each of these two samples were taken (column 3) and determinations were made of the calcium concentration in parts per million (column 4). The sums of squares (column 5) and the mean values (column 6) of the determinations in each leaf were calculated. The mean values in column 6 were used to determine the sums of squares (column 7) and the mean values (column 8) for the leaves within plants. The mean values in column 8 were use to determine the sums of squares within plants (column 9)

1	2	3	4	5	6	7	8	9
P	L	S	Ca%	SS(S in L)	⟨S in L⟩	SS(L in P)	⟨L in P⟩	SS(P)
1	1	1	3.28	0.0180	3.19			
		2	3.10					
	2	1	3.52	0.0008	3.50	0.2180	3.18	
		2	3.48					
	3	1	2.88	0.0032	2.84			
		2	2.80					
2	1	1	2.46	0.0002	2.45			
		2	2.44					
	2	1	1.87	0.0012	1.90	0.1542	2.18	
		2	1.92					
	3	1	2.19	0.0000	2.19			
		2	2.19					
3	1	1	2.77	0.0060	2.72			1.2601
		2	2.66					
	2	1	3.74	0.0450	3.59	0.6248	2.95	
		2	3.44					
	3	1	2.55	0.0000	2.55			
		2	2.55					
4	1	1	3.78	0.0041	3.83			
		2	3.87					
	2	1	4.07	0.0012	4.10	0.3181	3.74	
		2	4.12					
	3	1	3.31	0.0000	3.31			
		2	3.31					

where from each plant p_i three leaves were chosen so that l_{ij} is leaf j in plant i and from each leaf two samples were taken and used to measure the concentration of calcium so that s_{ijk} is kth determination of concentration of calcium in leaf j from plant i.

We can now calculate the sums of squares for samples in leaves, leaves in plants and plants. For the samples in leaves we simply add up the individual sums of squares given in column 5. For the leaves in plants we add up the sums of squares in column

Table 8.22 Sums of squares and mean squares for the data given in Table 8.23

Source	SS	d.f.	MS
Plants	7.5603	3	2.5201
Leaves in plants	2.6302	8	0.3288
Samples in leaves	0.0799	12	0.0067

7, but since each mean is the mean of two determinations, we multiply by two to find the total sum of squares for the leaves in plants. For the plants we take the sum of squares in column 9 and since each mean is the mean of six determinations, we multiply by six to find the total sum of squares for the plants. The results of these calculations are given in Table 8.22. To determine the mean squares, we now divide each sum of squares by the number of degrees of freedom. Since there are four plants, there are 3 degrees of freedom for the plants. Since there are three leaves within each of four plants, there are $(3 - 1) \times 4 = 8$ degrees of freedom for the leaves in plants. Since there are two samples within each of 3×4 plants, there are $(2 - 1) \times 12 = 12$ degrees of freedom for the samples in leaves.

Now we need to think carefully about just what each mean square estimates. The mean square for samples in leaves allows for the effects of the plants and the leaves, since each term is calculated after allowing for their effects, and gives us an estimate of σ_s^2, the population variance of the samples in each leaf. If neither the plants nor the leaves had any effect on the measurements, the mean squares for the leaves in plants would also give us an estimate of σ_s^2. Suppose, however, that the population variance of the calcium concentration among the leaves is σ_l^2. Then we add to the mean square for the leaves in plants $n_s \sigma_l^2$ where we have multiplied by n_s the number of samples within each leaf. The mean square for leaves in plants is therefore an estimate of $\sigma_s^2 + n_s \sigma_l^2$ (c.f. section 8.1.3). The variation among plants includes the variation among leaves so that we start with $\sigma_s^2 + n_s \sigma_l^2$. Now if the population variance of the calcium concentration among the plants is σ_p^2 we must add $n_s n_l \sigma_p^2$ where we multiply by $n_s n_l$ the number of samples within each plant. These results are summarized in Table 8.23. To test the null hypothesis that the plants have no effect, that is $\sigma_p^2 = 0$, we use the mean square for leaves within plants as our error term. To test the null hypothesis that the leaves have no effect, that is $\sigma_l^2 = 0$, we use the mean square for samples within leaves as our error term. From Table 8.23 we see that the variation due to the plants is significant at the 5% level while the variation due to the leaves is significant at the 0.1% level.

This experiment was performed to determine the overall variation in the amount of calcium in turnip plants. We want to know how much of the variation is due to plants, leaves and samples taken from each leaf. The difference between the mean squares for plants and for leaves in plants divided by $n_s n_l$ gives us 0.365 as our

Table 8.23 Analysis of variance table for the determination of calcium within leaves on turnip plants. The F ratio for plants is calculated as $2.5201/0.3288 = 7.66$ with 3 and 8 degrees of freedom. The F ratio for leaves is calculated as $0.3288/0.0067 = 49$ with 8 and 12 degrees of freedom

Source	d.f.	MS	Parameters estimated	F-ratio	*
Plants	3	2.5201	$\sigma_s^2 + n_s\sigma_1^2 + n_s n_1 \sigma_P^2$	7.66	*
Leaves in plants	8	0.3288	$\sigma_s^2 + n_s\sigma_1^2$	49	***
Samples in leaves	12	0.0067	σ_s^2		

best estimate of σ_P^2. The difference between the mean squares for leaves in plants and samples in leaves divided by n_s gives us 0.161 as our best estimate of σ_1^2. Finally, the mean square for samples gives us our best estimate of σ_s^2 as 0.0067. The standard deviations are then 0.60, 0.40 and 0.08, for plants, leaves and samples, respectively. The variation among leaves on a given plant is almost as great as the variation among plants, whereas the variation among determinations on a given leaf is much less.

With a nested design and random factors we are interested in the amount of variation within successive levels of the various factors and our analysis is aimed at estimating the contribution to the overall variation from the variation at each level. This is in contrast to a crossed design with fixed factors where we are interested in comparing the different levels of each factor using a multiple range test. It is important to remember that for random factors we tacitly assume that their effects are normally distributed.

8.5 SUMMARY

Most of the variables that we measure have associated uncertainties that are normally distributed or can be made so by a well-chosen transformation. Furthemore, if we design our experiments carefully, the sum of the squares of the deviations of all of our observations from the mean can be decomposed into contributions from each of the contributing factors. For these reasons the analysis of variance (which should, perhaps, be called the analysis of sums of squares) is central to the statistical analysis of experimental data.

ANOVA allows us to take the results of a complicated experiment with many factors, crossed and nested, fixed and random, some interacting with others, and identify those that contribute significantly to the overall variation in the data. We can then analyse the data further using multiple range tests for fixed factors or the added variance components for random factors. The design of our experiments can be as complicated as we choose although the analysis of variance will then be correspondingly complicated.

The most complicated experimental design can be expressed as combinations of crossed and nested factors. We say that B is nested in A if the levels of factor B

differ among the levels of factor A. In Student's data on soporific drugs (section 8.1.6), if there had been 20 patients of whom ten received one drug and ten the other, the patients receiving the first drug would not have been the same as the patients receiving the second drug so that patients would have been nested within drugs. We then analyse the data using an unpaired t test or a single-factor ANOVA. The turnip data discussed in section 8.4.6 is an example of a purely nested design, with determinations nested in leaves, leaves nested in plants. If a factor B is nested within a factor A, factor B must be a random factor and we are then interested in using the ANOVA to determine the variance due to factor B.

We say that factor B is crossed with factor A if each level of factor B occurs in combination with each level of factor A. In Student's data on soporific drugs (section 8.4.4), we actually had ten patients, each of whom received both drugs. Since each patient was given each drug, the patients were crossed with the drugs. We were then able to analyse the data using a paired t test or a two-factor ANOVA. Although factor B was now crossed with factor A, factor B (patients) was still a random factor while factor A (drugs) was a fixed factor. If we were interested only in these ten people and never had any intention of giving the drug to anyone else, the patients would also have been a fixed factor. We are, of course, testing the drugs on these ten patients in the hope that if the drugs are effective we will then be able to give them to other patients. Our ten patients are therefore a random sample from a larger population of patients. The data on the weight gained by male and female rats fed on fresh and on rancid lard is another example of a purely crossed design. In this case, sex and lard freshness are both fixed factors.

The analysis of our data depends on whether the factors are nested or crossed, random or fixed. Especially in agricultural trials, field crops and treatments may be divided up in quite complicated ways and one needs to think carefully about the analysis. As always, the best defence against error and confusion is to think carefully about the biology and to think about the relationship between the particular experiment that you have performed and the broader context in which you hope to apply your results.

In this chapter our response variables have generally been quantitative (hours of sleep, number of ticks, weight gained), while our predictor variables have been qualitative (drugs, rabbits, sex and lard freshness). In Chapter 7 we touched on non-parametric tests, in the analysis of data for which the response variables are also qualitative. There is no reason to restrict ourselves to qualitative predictor factors, however, and in the next chapter we will consider the analysis of data in which both the response and the predictor variables are measured on quantitative scales.

8.6 EXERCISES

1. In a single factor ANOVA for an experiment on four rabbits, ten ticks were removed from each rabbit and weighed. We are interested in the variation between the average weights of the ticks on the four rabbits. How many degrees of freedom would you use to calculate the within-treatments sums of squares? How many degrees of freedom

Table 8.24 The number of male *G. pallidipes* tsetse flies caught by a biconical, *b*, an epsilon, *e*, and a Nguruman, *n*, trap. These data were collected as part of the experiment described in section 8.4.5

Site	Day 1	Day 2	Day 3
1	n675	e 192	b292
2	b193	n114	e 245
3	e 395	b109	n507
4	e 132	b 39	n214
5	n506	e 94	b191
6	b 89	n105	e 210
7	b 25	n 46	e 43
8	e 34	b 18	n160
9	n 49	e 40	b 55

Table 8.25 Analysis of variance table for the combined Latin rectangles given in Tables 8.17 and 8.24

Source of variation	Sums of squares	Degrees of freedom	Mean square	F ratio	Significance level, P
Sex	0.050	1	0.050	2.31	0.14
Trap	2.465	2	1.232	56.3	0.0000
Day	0.698	2	0.349	15.9	0.0000
Site	4.116	8	0.514	23.48	0.0000
Sex × Trap	0.270	2	0.135	3.19	0.0060
Sex × Day	0.226	2	0.113	6.18	0.0122
Sex × Site	0.342	8	0.043	5.17	0.0912
Residual	0.613	28	0.218	1.95	

would you use to calculate the beween-treatments sums of squares? How many degrees of freedom would there be in the total sums of squares?

2. Using the data in Table 8.16 estimate the added variance component and hence the population standard deviation in the amount of sleep gained due to the variability among patients.

3. How many degrees of freedom are there for each factor and for the error term in a 4 × 4 Latin square? How many measurements would you make? How many degrees of freedom are there for each factor and the error term in a 4 × 4 factorial experiment? How many measurements would you make?

4. Use a series of F tests to decide if any of the standard deviations given in Table 8.7 differ significantly.

5. The data for male flies given in Table 8.24 were collected at the same time as the data for female flies given in Table 8.17. Using a computer carry out an analysis of variance treating the three squares as one Latin rectangle.

6. The data for the number of female and male flies caught by the three traps given in Tables 8.19 and 8.24 can be combined into a single table with sex as an additional factor. Table 8.25 gives the results of the ANOVA on the combined data. Interpret the significance of the interactions.

9

Regression

Nature only uses the longest threads to weave her patterns, so each small piece of her fabric reveals the organization of the entire tapestry.

R. Feynman (1980)

In the previous chapter we used one or more **independent** variables (drugs, rabbits, lard freshness) to explain the variation in a **dependent** variable (hours of sleep, scutum width of larvae, weight gained by rats). When the independent variables are categorical, the analysis of variance provides a formal structure that enables us to assess the contribution that each of them makes to the variation in the dependent variable. Consider, however, the control of armyworm using *Bacillus thuringiensis* (B.t.) discussed in section 1.1.4. Brownbridge (1988) was concerned to determine the biocidal effect of B.t. on armyworms. In his experiments, the results of which are given in Table 9.1, the control areas were not sprayed, while 0.5, 1.0 and 2.0% B.t. suspended in water was applied for the three treatments. Estimates were made of the number of armyworms surviving under each treatment on five successive days.

We could analyse these data using a two-factor ANOVA with days and treatments as the factors and our model would be

$$n_{ij} = m + a_i + b_j + \varepsilon_{ij},\qquad 9.1$$

where a_i accounts for the effect of the ith treatment and b_j for the effect of the jth day on the number of armyworms. We would then find that the average number of armyworms declined over time under all three treatments and that fewer armyworms survive under treatment C than under the control, for example.

Let us try a different approach. We really want to know if the number of armyworms declines at different rates under different treatments. We see from Fig. 9.1 that the logarithm of the number of armyworms decreases more or less linearly with time in all four cases. The equation for a straight line is $y = a + bx$ and this suggests a model of the form

$$y_i = ln(n_i) = a + bt_i + \varepsilon_i,\qquad 9.2$$

where a is the intercept on the $ln(n)$ axis, b is the slope of the line, t_i is the time at which the ith observation was made and ε_i is the residual term that gives us the

Table 9.1 The number of armyworms counted on each of five successive days under each of four treatments. *Cont.* is the unsprayed control sample, *A*, *B* and *C* are the areas sprayed with 0.5%, 1.0% and 2.0% *Bacillus thuringiensis* suspended in water, respectively. The last four columns give the natural logarithms of the numbers of armyworms

Day	Cont.	A	B	C	ln(Cont.)	ln(A)	ln(B)	ln(C)
0	497	320	294	295	6.209	5.768	5.684	5.687
1	463	203	213	93	6.138	5.313	5.361	4.533
2	506	155	118	33	6.227	5.043	4.771	3.497
3	487	125	124	12	6.188	4.828	4.820	2.485
4	480	101	111	4	6.174	4.615	4.710	1.386

deviation from the line for the ith observation. (For the moment we will apply Equation 9.2 separately to each treatment.)

The model given in Equation 9.2 is attractive for several reasons. If we fit separate lines to the data for each treatment, then in each case we have 5 degrees of freedom and lose 2 for the model (the estimates of the slope and the intercept) leaving us with 3 degrees of freedom for the error term. We can then compare our estimates of the slopes to see if the various treatments differ in their effect on the armyworms. Once we have found values of a and b that give us the best fit to each set of data, we can then choose any value for t and use Equation 9.2 to calculate the corresponding value of $ln(n)$. The problem is to find values of a and b, the coefficients in the fit and this is what regression does for us.

Before we find out how to determine the values of a and b, let us ask why the

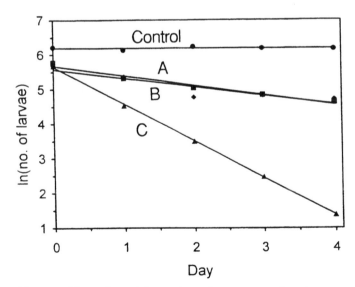

Figure 9.1 The natural logarithms of the numbers of armyworms plotted against time. Under the control no application of the biocide was made while the three treatments *A*, *B* and *C* were made with 0.5, 1.0 and 2.0% *B.t* suspended in water, respectively.

data fall on a straight line when the logarithms of the numbers are plotted against time. In the experiment on armyworms, we might expect the same proportion of those surviving to die on each successive day. To express this idea formally, we write

$$dN(t)/dt = rN(t),\qquad\qquad 9.3$$

where dN/dt is the rate of change of N per unit time and the proportionality constant r gives the growth or decay rate of the population. (For readers unfamiliar with calculus, Equation 9.3 says that if we take a very small interval of time, dt, then dN, the change in the number of armyworms in this small interval of time, will also be very small. If we divide the one small number by the other, their ratio, which gives the number dying per unit time, is proportional to N, the number present, and the proportionality constant is r. Learning calculus enables you to derive Equation 9.4 from Equation 9.3.)

Integrating Equation 9.3 gives

$$N(t) = N(0)e^{rt},\qquad\qquad 9.4$$

where $N(t)$ is the number of armyworms at time t, $N(0)$ is the number at $t = 0$, and r is the rate of increase or decrease, that is, the fractional change in numbers per unit time. If $r < 0$, Equation 9.4 describes an exponential decline, if $r > 0$, an exponential growth. (For a discussion of the dynamics of single populations, see May, 1981.) Taking logarithms of both sides of Equation 9.4, we have

$$ln[N(t)] = ln[N(0)] + rt,\qquad\qquad 9.5$$

so that if we plot the logarithm of the numbers sampled on each day against time, we expect the numbers to lie on a straight line as they do in Figure 9.1. The slope of the line gives the mortality rate and the intercept on the vertical axis gives the number originally present so that both have a direct biological interpretation.

The word 'regression' has an odd etymology. In 1889 Francis Galton formulated his 'law of universal regression' (Galton, 1889), which stated that 'each peculiarity in a man is shared by his kinsman, but on the average in less degree'. To illustrate this, Figure 9.2 shows a plot of the heights of sons against the heights of their fathers from records of more than a thousand families collected by Pearson and Lee (1903). The equation of the line shown in Fig. 9.2 is

$$y = 37.75 + 0.457x.\qquad\qquad 9.6$$

The average height of the fathers is 67.0 inches. Using Equation 9.6 to calculate the corresponding height of the sons gives 68.4 inches, so that on average the sons were 1.4 inches taller than their fathers. However, we can use Equation 9.6 to show that a father who is 6 inches above the average height (for fathers) produces, on average, a son who is only 2.7 inches above average height (for sons) and a father who is 6 inches below the average height produces, on average, a son who is only 2.7 inches below the average height (for sons). Galton therefore argued that there was a 'regression' or 'going back' towards the average. Today we use the word 'regression' to mean fitting a line to a plot of one variable against another; the use of the

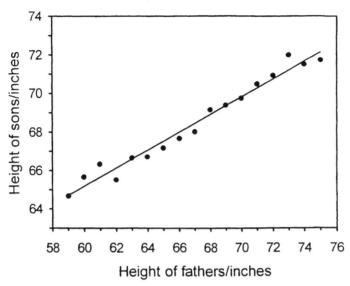

Figure 9.2 Plot of the heights of sons against the heights of their fathers from 1078 families. Each point is the average height of all the sons whose fathers heights are 59, 60, 61, ... inches.

word regression no longer implies Galton's law. (For a discussion of the genetics of inheritance based on Galton's data, see Roughgarden, 1979, p. 135.)

Once we have decided on the line that gives the best fit to a set of data, we can **interpolate** the data and use the fitted line to estimate values of the **dependent** variable (son's height in Fig. 9.2) for any value of the **independent** variable (father's height in Fig. 9.2). We can also **extrapolate** the fitted line beyond the extremes of the data and make the prediction that a father who is 78 inches tall is likely to produce sons who are 74 inches tall, even though the tallest father whose height was actually measured was only 75 inches.

9.1 DEFINING THE FIT

When we carry out a regression we want to find the line that gives us the best fit to our data, but we cannot do this until we decide what we mean by 'best'. We have already seen in section 4.1 that the location of a distribution can be defined in terms of the mean, the median or the mode, and which one is 'best' depends on what we intend to do with our data. The mean has certain rather attractive properties, in particular, that it is the number about which the sum of the squares of the residual deviations is a minimum. (We can also show that if the numbers come from a normal distribution whose true mean is μ, then the sample mean m is the estimate of μ that has the smallest variance.)

Since the mean has this rather desirable property, it makes sense to try a similar approach in regression and define the 'best fit' as the line that minimizes the sum of the squares of the residual deviations from the fit. Fig. 9.3 shows a line drawn through four points. For each point x_i we have the observed value of y, y_i, and the value of

Regression

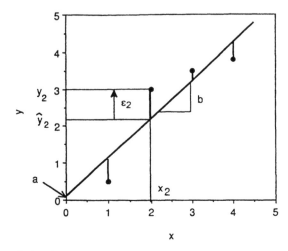

Figure 9.3 A straight line has been drawn through four points. Each residual ε_i is the difference between the observed value of the dependent variable and the value read from the fitted line. The best fit line minimizes the sum of the squares of the residuals. a is the intercept on the y axis when x is zero, b is the slope of the line.

y, read from the line, which we will call $\hat{y}_i$. The difference between these two values of y is the residual from the fit, which we will call ε_i, so that

$$\varepsilon_i = y_i - \hat{y}_i. \tag{9.7}$$

Since $\hat{y}_i$ is the value of y read from the fitted curve, we can write

$$\varepsilon_i = \hat{y}_i - a - bx_i. \tag{9.8}$$

We now want to find the values of a and b that minimize the sums of squares of the residuals, $\sum \varepsilon_i^2$, and it is straightforward to show (Freund, 1972, p. 455) (with the aid of calculus) that these are

$$a = \frac{\Sigma y \Sigma x^2 - \Sigma xy \Sigma x}{n \Sigma x^2 - (\Sigma x)^2} \tag{9.9}$$

$$b = \frac{n \Sigma xy - \Sigma x \Sigma y}{n \Sigma x^2 - (\Sigma x)^2}. \tag{9.10}$$

(I have dropped the subscript i from each x and y in Equations 9.9 and 9.10 to make them easier to write; if you wish, you may put the subscripts in.) The derivation of Equations 9.9 and 9.10 assumes that the only errors are in y and that the errors in x are zero or at least negligibly small. The derivation also assumes that all of the ε_i are normally distributed with mean zero and a constant variance.

We can now carry out the regression of the number of armyworms against time under treatment C. Using Equations 9.9 and 9.10, the logarithm of the number of armyworms surviving t hours after applying the treatment is

$$\ln[N(t)] = 5.647 - 1.065t. \tag{9.11}$$

The best estimate of the number of larvae on day 0 is $e^{5.647} = 284$ and the larvae are dying at a rate of $1.065/24 = 0.044$, or 4.4%, per hour.

9.2 ANOVA

When we carry out a regression, we want not only to find the best-fit line but also to know how good the fit is. There are several ways to do this. One way is to carry out an ANOVA on the fitted line. Just as before, we begin by subtracting the mean since we are usually concerned only with deviations from the mean and our null hypothesis is that the 'true' values of y are constant. In Brownbridge's study of the African armyworm, under treatment C, we start by calculating the total sums of squares about the mean, SS_T, and then break this down into a sum of squares due to the fit, SS_F, and a sum of squares due to the residuals from the fit, SS_R.

Using y_i for the measured points, $\hat{y}_i$ for the corresponding points on the fitted line, **after subtracting the mean**, and ε_i for the residuals, we have

$$y_i = \hat{y}_i + \varepsilon_i. \tag{9.12}$$

Squaring Equation 9.12 and summing over i

$$\Sigma y_i^2 = \Sigma \hat{y}_i^2 + 2\Sigma \hat{y}_i \varepsilon_i + \Sigma \varepsilon_i^2. \tag{9.13}$$

The left-hand side is the total sum of squares, SS_T, the first term on the right-hand side is the sum of squares from the fit, SS_F, and the last term on the right-hand side is the sum of squares due to the residuals, SS_R. In the Appendix (section 9.9.1) I show that the middle term on the right-hand side of Equation 9.13 is zero, but since the residuals ε_i are as likely to be positive as negative, it seems reasonable that the sum over $\hat{y}_i \varepsilon_i$ should be zero, and Equation 9.13 may be written

$$SS_T = SS_F + SS_R. \tag{9.14}$$

Since we have calculated SS_T as the sum of squares about the mean, we have lost 1 degree of freedom, and

$$d.f._T = n - 1. \tag{9.15}$$

We have then used up one more degree of freedom in setting the slope of the line, so that for the fitted line we have

$$d.f._F = 1, \tag{9.16}$$

leaving $n - 2$ degrees of freedom for the residuals, so that

$$d.f._R = n - 2. \tag{9.17}$$

Now we proceed just as in a conventional ANOVA: dividing each sum of squares by the appropriate number of degrees of freedom gives us the mean squares. If the

Table 9.2 ANOVA for the line corresponding to treatment C in Figure 9.1

Source	Sum of squares	Degrees of freedom	Mean square	F-ratio	Significance level, P	*
Model	11.3402	1	11.3402	6140	0.000 005	***
Error	0.0055	3	0.0019			
Total	11.3457	4			$R^2 = 99.95\%$	

mean square for the fit, MS_F, is large while that for the residuals, MS_R, is small, the fit accounts for much of the total variance and is significant. If the reverse is the case, the fit does not account for much of the variance and the fit is not significant. The F ratio is calculated as

$$F = MS_F/MS_R \qquad\qquad 9.18$$

and we test our calculated F ratio against $F_{1,n-2}$.

The ANOVA for the fit to the armyworm data for treatment C is given in Table 9.2 and the value of F is 6140, so that the fit is highly significant. Table 9.2 also gives a parameter, R^2, which is the proportion of the variance explained by the fit: in this case $11.3402/11.3457 = 0.9995$. (Strictly, R^2 is the proportion of the total sum of squares accounted for by the model, but it is referred to as the proportion of the total variance accounted for by the model.) Since 99.95% of the variance in the data is explained by the fit, it is not surprising that the fit is highly significant.

9.3 ERROR ESTIMATES AND CONFIDENCE LIMITS

The analysis of variance applied to a regression enables us to determine the statistical significance of a line fitted to a set of data. However, we would like to determine error limits for our estimates of the parameters in our fit, especially as we hope to interpret the parameters biologically. We would also like to place confidence limits on our fitted line so that we can say that the true line lies within some range about the fitted line with an appropriate degree of certainty.

9.3.1 Parameters

Equations 9.9 and 9.10 enable us to calculate the regression coefficients for a straight line fit to a set of data. The derivation of these equations is based on the assumption that the only errors are those associated with the dependent variable, y, and that there are no errors associated with the determination of the independent variable, x. If this is so, we can calculate the errors in a and b corresponding to the errors in y (Snedecor and Cochran, 1989, p. 175). The standard deviation of the residuals is calculated in the usual way, as

$$s_p^2 = \Sigma \varepsilon_1^2/(n-2), \qquad\qquad 9.19$$

Table 9.3 Regression of the natural logarithm of the number of armyworm larvae against time for treatment C. The table gives the least squares estimate of the slope and intercept together with its standard deviation, the t statistics and the significance level. The slope has units of days^{-1}

Parameter	Value	s	t	P	*
Intercept	5.647	0.043	131	<0.00001	***
Slope	− 1.065	0.014	− 78	<0.00001	***

where the sum is from $i = 1$ to n and we have divided by the number of degrees of freedom that remain after fitting two parameters, the intercept and the slope. The standard deviation of the mean value of the dependent variable, y, is then

$$s_m^2 = s_p^2/n, \qquad 9.20$$

the usual expression for the standard error of a mean. The standard deviation of the slope, b, is (Appendix, section 9.9.2)

$$s_b^2 = s_p^2/\Sigma (x_i - \bar{x})^2. \qquad 9.21$$

For the regression of the number of armyworms against time under treatment C, Table 9.3 gives the standard deviations of the intercept and the slope and if we do a t test on each we find that both differ significantly from zero at the 0.1% level. Furthermore, if we multiply the standard deviation of the slope by the critical value for a two-tailed t test at the 5% significance level, we obtain 95% confidence limits for the slope as $− 1.065 \pm 0.014 \times 3.18$, so that we can assert that the rate of decline of the number of armyworms lies between $− 1.020$ and $− 1.110$/day with 95% confidence.

In our armyworm experiments we want to know how the rate of decline varies between treatments to see which, if any, reduces their number significantly. If we fit straight lines to each of the four sets of data given in Table 9.1, after taking logarithms, we obtain the rates of decline given in Table 9.4. We test each t statistic using a one-tailed test since we are interested in the treatment only if it produces a decline in the numbers over time. In each case there are 3 degrees of freedom since each line is fitted to five points. The slope of the control curve is not significantly less than zero while all three treatments give significant rates of decline, B at the 5% level, A at the 1% level and C at the 0.1% significance level.

Table 9.4 The daily rates of decline in the number of armyworms under the four treatments described in Table 9.1 calculated from the slopes of straight lines fitted to the data

Treatment	Slope	s	t	*	
Control	− 0.0019	0.0124	0.15	ns	a
B	− 0.2489	0.0642	3.88	*	b
A	− 0.2791	0.0284	9.83	**	b
C	− 1.0649	0.0136	78.3	***	c

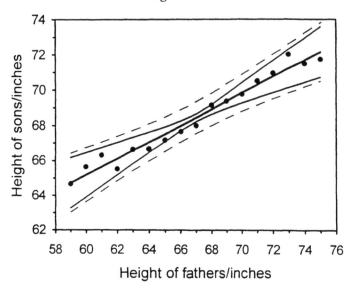

Figure 9.4 The heavy solid line is the best fit straight line to the measured points. The thin solid line is a 95% confidence band for the fitted line. The dashed line is a 95% confidence interval for the individual points so that, on average, 19 out of 20 points should lie between the dashed lines.

These experiments were conducted in order to determine the concentration of B.t. to use in armyworm control. To make comparisons between the various treatments, we calculate the difference between pairs of slopes divided by the pooled standard deviation of all the slopes, after checking that the variances of the slopes do not differ significantly. Since we do not know in advance whether the difference between two slopes will be positive or negative, we use two-tailed tests. Since each slope has 3 degrees of freedom, each comparison has 6 degrees of freedom. Since we can make $4(4-1)/2 = 6$ comparisons, we need to use a significance level of $(5/6)\%$ to ensure that the significance level for all comparisons is at least 5%. When this is done, treatment B does not differ significantly from treatment A, but all other pairs differ significantly at the 5% level, as indicated in Table 9.4. In particular, treatment C is substantially better than treatments A or B.

9.3.2 Lines

Having fitted a line to a set of data, we can also calculate a confidence band for the line itself which will enable us to set limits within which the 'true' line lies with a suitable degree of confidence. The uncertainty in the estimate of the line can be regarded as arising from uncertainty in the estimate of the mean value and uncertainty in the estimate of the slope. The expression for the standard deviation of the fitted line at point x is (Appendix, section 9.9.3)

$$s_1^2 = s_m^2 + s_b^2 \times (x - \bar{x})^2. \qquad 9.22$$

The standard deviation of the fitted line at the mean ($x = \bar{x}$) is simply the standard deviation of the mean of y, since the line passes through the point $(\bar{x}, \bar{y})$, and as x deviates from the mean in either direction, the standard deviation increases because of the uncertainty in the slope of the line. Again we need to multiply by t_{n-2} for a two-tailed test at the 5% significance level to obtain the corresponding 95% confidence band.

Figure 9.4 shows the data plotted in Fig. 9.2 with a 95% confidence band indicated by the thin solid lines. In regression tables, such as Table 9.3, we usually specify the standard error of the intercept on the y axis and this may be obtained by putting $x = 0$ in Equation 9.22.

9.3.3 Points

We would also like to set limits on the region within which the individual points should lie. This will enable us to detect outliers in our data. Equation 9.19 gives the variance of the residuals from the fitted line and Equation 9.22 the variance of the fitted line. To calculate a 95% confidence band for the individual points, we simply add these two variances, take the square root and multiply by t_{n-2}. For our graph of the heights of sons against the heights of their fathers, this gives the dashed line in Fig. 9.4. Approximately 19 out of 20 points should lie within the dashed lines.

9.4 RESIDUALS

This book began with a discussion of exponential growth and its influence on Darwin. Exponential growth is one of the most important ideas in ecology, so let us restate it: whenever the change in the number of objects is proportional to the number already present, the increase or decrease in the number follows an exponential curve. If 10 plants produce 1 new plant per week (on average), 20 will produce 2, 50 will produce 5 and so on, and the change in the number per week (1, 2 or 5) is always 1/10th of the number already present (10, 20, 50). For the same reason, if the birth rate exceeds the death rate by a constant amount, human populations increase exponentially; if the rate of inflation is constant, the value of your money falls exponentially; because the number of radioactive atoms that decay in a given time is proportional to the number that have not yet decayed, the intensity of radiation emitted by radioactive isotopes decreases exponentially with time.

Darwin realized that exponential growth cannot go on forever (remember our calculation on the potential growth of tsetse fly populations in Chapter 1). All natural populations are constrained by density dependent factors that limit further growth at a level referred to as the carrying capacity of the habitat for that population. The convergence to the carrying capacity is itself often exponential.

To illustrate these ideas, consider the experiment of Ashby and Oxley (1935) who observed the growth of duckweed under controlled conditions in their laboratory and counted the number of fronds at the beginning of each day. Their data are given in Table 9.5 and plotted in Fig. 9.5. If the growth is exponential, a plot of $ln[N(t)]$ against t should give a straight line with a slope of r and intercept on the vertical

Table 9.5 The number of duckweed fronds in a beaker kept at 24°C on each of 14 successive days

Day	1	2	3	4	5	6	7
Number	100	127	171	233	323	452	654
Day	8	9	10	11	12	13	14
Number	918	1406	2150	2800	4140	5760	8250

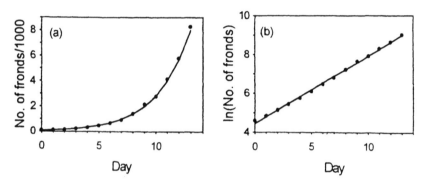

Figure 9.5 (a) The number of duckweed fronds and (b) the logarithm of the number of duckweed fronds plotted against time.

axis equal to the logarithm of the number of fronds on day 0 (Equation 9.5). Figure 9.5b shows that this is indeed the case and the parameters from the fit are given in Table 9.6. The slope of the curve is 0.349 ± 0.005/day so that the growth rate is $34.9 \pm 0.5\%$ per day. With this value of r, $e^{rt} = 2$ when $t = 1.99$ days, so that the number of fronds doubles every 2 days.

Although the fit in Fig. 9.5b looks good, it is always worth plotting the residuals as a check. If we do this we obtain the result shown in Fig. 9.6, from which it is clear that there is something odd going on at the beginning of the experiment. One possibility might be that we need a certain number of fronds to start the process. If we let the growth rate be proportional to the number of fronds above a threshold value, say M, we can replace Equation 9.3 by

$$dN(t)/dt = r(N(t) - M).\qquad\qquad 9.23$$

Integrating Equation 9.23 gives

$$N(t) - M = N(0)e^{rt}\qquad\qquad 9.24$$

Table 9.6 Parameters for the regression line shown in Figure 9.5b. The slope of the line is given in unit of fronds/day

Parameter	Value	s	t	P	*
Intercept	4.107	0.039	105	< 0.0001	***
Slope	0.349	0.0046	76	< 0.0001	***

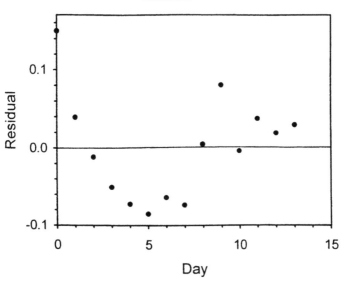

Figure 9.6 The residuals from the plot shown in Figure 9.5b.

so that

$$ln[N(t) - M] = ln[N(0)] - rt. \qquad 9.25$$

Since we do not know M, we cannot plot the data in a form that will give us a straight line directly: what we can do is to use an **iterative** procedure in which we repeatedly guess different values of M until we find the value that gives us the best fit. We do not of course carry out the iteration blindly: if we change M and the fit deteriorates, we know that we are changing it in the wrong direction. One way to decide when we have found the best value of M is to do an ANOVA on each fit and find the F ratio for each value of M. The value of M that maximizes the F ratio turns out to be 33. If we use this value and plot $ln(N(t) - 33)$ against time, we obtain the result shown in Fig. 9.7; the parameters from the fit are given in Table 9.7. The slope of the regression line gives a new estimate of the growth

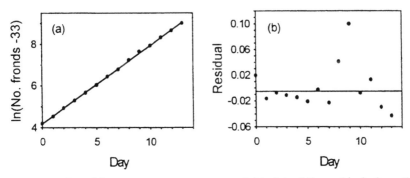

Figure 9.7 (a) Plot of $ln[N(t) - 33]$ against time and (b) plot of the residuals from the fit against time.

Table 9.7 Parameters from the regression of $ln[N(t) - 33]$ against time and the corresponding ANOVA table

	Parameter	Value	s	t	P	*
	Intercept	3.81	0.021	181	< 0.0001	***
	Slope	0.375	0.0025	151	< 0.0001	***
Source of variation	Sums of squares	Degrees of freedom	Mean square	F-ratio	Significance level, P	*
Model	31.95	1	31.95	23 059	< 0.0001	***
Error	0.017	12	0.0014			
Total	31.967	13			$R^2 = 99.95\%$	

rate of 0.375 ± 0.0025 fronds/day, slightly lower than before, but with a threshold of 33 fronds for the process to start. Plotting the residuals in Fig. 9.7b we see that the overall fit is considerably better than before, although the number of fronds on day 9 is still some way off.

We now go back to the biology: why are 33 fronds needed for the growth to start? What would happen in a smaller or a larger beaker? Eventually the numbers must level off, but at what number of fronds? Can we deduce anything useful from the fact that the doubling time is about 2 days? The statistics, if calculated carefully, will raise questions, but the answers can be found only in the biology.

9.5 CORRELATION

We have seen in the previous sections how to obtain the best estimate of one variable based on the values of another. However both variables are often subject to error and neither is more fundamental than the other. In such situations we might be concerned with the degree of correlation between the two variables rather than the dependence of one on the other.

In section 4.2.1 we defined the variance of a set of numbers as

$$V = \Sigma(x_i - \bar{x})^2/n. \qquad 9.26$$

Furthermore, we have repeatedly spoken about variables being 'statistically independent'. To make this concept of statistical independence more precise, we define a new parameter, which we call the **covariance**, as

$$C(x, y) = \Sigma(x_i - \bar{x})(y_i - \bar{y})/n. \qquad 9.27$$

The analogy with the definition of the variance is clear since $C(x, x) = V(x)$, but we need to convince ourselves that the definition is sensible.

In Equation 9.27 we have, as usual, subtracted the mean of each set of numbers. If x and y are either both positive or both negative, after subtracting their means, their product will contribute positively to the covariance, so that if they go up and down together, the covariance will be positive. In the same way we can see that if one goes up when the other goes down, or vice versa, they will contribute negatively to the covariance. We can restate this in terms of our regression and say that if the regression line has a positive slope, the covariance will be positive, and if the regression line has a negative slope, the covariance will be negative. We can now write down a more complete expression for the variance of a sum:

$$V(x + y) = V(x) + V(y) + 2C(x, y) \qquad 9.28$$

and Equation 4.8, which we have used to evaluate the variance of a sum, is correct only when the covariance of the two variables is zero.

Just as the standard deviation is often more useful than the variance because it has the same units as the original measurement, it turns out that a different measure of association, the **correlation coefficient**, is often more useful than the covariance. Taking the square root of the covariance does not help us much because the units of the two variables may be different, and if they were, say, metres and grams, we would have something measured in $metres^{1/2} grams^{1/2}$. Indeed, since we have already accounted for the variance of each measurement through their separate standard deviations, the most useful measure involves dividing the covariance by the standard deviation of each variable so that the **correlation coefficient**, r, is

$$r = C(x, y)/[V(x) \times V(y)]^{1/2}. \qquad 9.29$$

If the points lie exactly along a straight line with positive slope, the correlation coefficient is $+1$, and if they all lie exactly along a line with negative slope, the correlation coefficient is -1. If there is no correlation between the two variables, the correlation coefficient is 0.

The correlation coefficient can be illustrated using the heights of 11 pairs of brothers and sisters (Pearson and Lee, 1903) given in Table 9.8 and plotted in Fig. 9.8. Calculating the correlation coefficient using Equation 9.29 gives

$$r = 0.553. \qquad 9.30$$

Since the heights of the brothers and sisters clearly tend to go up and down together, the correlation coefficient is positive but the relationship is not very close and the correlation coefficient is much less than 1.

When we calculate a correlation coefficient, or perform a regression or an ANOVA we are looking for relationships between two sets of numbers. In Fig. 9.8 the line marked (a) is a regression of the heights of the brothers against the heights of their sisters. However, if we carry out a regression of the heights of the sisters against the

Table 9.8 Heights, in inches, of 11 brothers and sisters, each pair belonging to a different family

Family	1	2	3	4	5	6	7	8	9	10	11
Brother	71	68	66	67	70	71	70	73	72	65	66
Sister	69	64	65	63	65	62	65	64	66	59	62

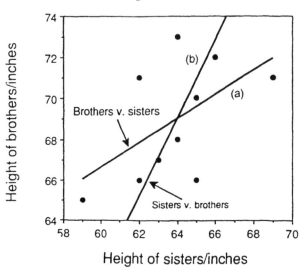

Figure 9.8 The heights of sisters plotted against the heights of their brothers using the data in Table 9.8. The two lines are regressions of brothers against sisters and sisters against brothers.

heights of their brothers and plot it on the same graph, we obtain the line marked (b) in Fig. 9.8. The reason that we can obtain two quite different regressions for the same set of data is because each regression is based on the assumption that the errors are only in the dependent variable and that the independent variable is measured without error. If the variability in the data arose only from the heights of the boys, line (a) would be correct; if variability in the data arose only from the heights of the girls, line (b) would be correct. Consequently, line (a) gives the best prediction of the height of a brother given the height of his sister, and line (b) gives the best prediction of the height of a sister given the height of her brother. If we are not interested in using heights of sisters to predict heights of brothers, or vice versa, but rather are interested in the degree of association between the two, neither regression is then 'correct' since both the boys' and the girls' heights are subject to random variations. In this case the advantage of the correlation coefficient is that it is not biased in favour of either. In fact, it is easy to show that the correlation coefficient is the geometric mean of the slopes of the two regressions lines, so that the correlation coefficient measuring the degree of association between the height of brothers and sisters is

$$r^2 = b_1 b_2,$$ 9.31

where b_1 is the slope of the regression of brothers against sisters and b_2 is the slope of the regression of sisters against brothers.

Finally, we can make a connection between the correlation coefficient and the ANOVA of our regression. If the correlation between two variables is perfect, the correlation coefficient should be ± 1. But the regression line should then explain all of the variance in the data, and the parameter, which we have called R^2 in Tables 9.2

and 9.7, should also be 1. If there is no correlation between the two variables at all, the correlation coefficient should be zero. But the regression should then explain none of the variance in the data and R^2 should be 0. In fact, it is not difficult to show that R and r are one and the same, so that the square of the correlation coefficient is the proportion of the total variance explained by the fit.

9.6 POLYNOMIAL REGRESSION

When we carry out a linear regression, we assume that the points lie along a straight line although we sometimes need to transform the data, perhaps by taking logarithms to achieve a linear relationship. If we cannot transform the data to lie along a straight line, we must use a more complicated function. One way to proceed is to fit a polynomial, which could be a quadratic function (a parabola), a cubic function or a still higher order polynomial. Consider, for example, the data measured by Crooke and Knight (1965) in a study of the relation between the age of oat plants and their content of nickel and iron when grown in sand cultures. Like other heavy metals, nickel supplied in excess to plants affects iron metabolism causing, in oat plants, longitudinal white necrotic striping of the leaves accompanied by an induced iron deficiency chlorosis.

As part of their study, Crooke and Knight measured the relative absorption of nickel and iron from a controlled nutrient solution supplying iron at 1.2 ppm and nickel at 2.5 ppm, with the results shown in Table 9.9 and plotted in Fig. 9.9. The data lie along a curve and if we decide to fit a parabola our model equation is

$$y_i = a + bx_i + cx_i^2 + \varepsilon_i. \tag{9.32}$$

As in the case of linear regression, we can again find expressions for the values of a, b and c that minimize the sums of squares of the residuals about the fitted line (Bliss, 1967, p. 37); the resulting values are given in Table 9.10 and the line is drawn in Fig. 9.9. The constant term is not significantly different from zero while both the

Table 9.9 The ratio of the amount of nickel to the amount of iron in oat plants as a function of age

Age/days	Ni/Fe	Age/days	Ni/Fe	Age/days	Ni/Fe
4	0.32	38	1.40	51	1.32
9	0.41	39	1.95	52	1.52
14	0.79	41	1.51	55	1.05
18	0.86	42	1.81	56	1.70
22	1.28	69	0.82	58	1.22
26	1.28	73	0.95	59	1.90
29	1.48	44	1.53	62	0.85
33	1.15	45	1.30	63	1.05
34	1.47	48	1.00	64	0.87
37	1.67	49	1.50	66	1.08

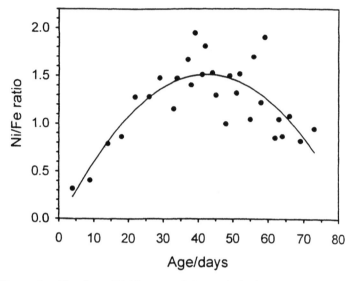

Figure 9.9 The data of Table 9.9 and the parabola that gives the best fit.

Table 9.10 Parameters for the regression line shown in Figure 9.9 and the ANOVA table. a is the intercept on the y axis, b and c are the coefficients of the linear and quadratic terms in the fit

Parameter	Value	s	t	P	*
a	-0.047	0.197	-0.240	0.812	ns
b	0.0735	0.01048	7.021	0.0000	***
c	-0.000866	0.000129	-6.7358	0.0000	***

Source	Sum of squares	Degrees of freedom	Mean square	F-ratio	Significance level, P	*
Model	3.04	2	1.52	24.7	0.0000	***
Error	1.66	27	0.0616			
Total	4.71	29			$R^2 = 65\%$	

linear and the quadratic terms are significant at the 0.1% level. From Fig. 9.9 we see that the ratio of nickel to iron increases up to age 42 days and then decreases as the trees age. We can again carry out an analysis of variance on the fitted line, from which we find that the model explains 65% of the variance in the data.

In our polynomial regression, the estimate of each coefficient depends on the inclusion of the others. If we fit a constant we will obtain an estimate of a. If we fit a straight line we will obtain an estimate of both a and b but we will find that the estimate of a has changed. If we fit a parabola we will obtain an estimate of a, b, and c but we will find that the estimates of a and b have changed. Formally we say that the constant, x, and x^2 are not orthogonal. The ANOVA tells us that the overall fit using all three terms is highly significant. The regression analysis gives us the significance level of each coefficient, after allowing for the other two. We should therefore consider first the significance of the quadratic term, c, in Table 9.10. If this is significant, we should keep all three terms. If it is not, we should drop it and repeat the exercise for the first two terms. We should drop terms of lower power first only if we have a sound biological reason for doing so. For example, it may be that we know that the graph must pass through the origin of our coordinate system, in which case we could drop the constant term.

9.7 MULTIPLE REGRESSION

Very often we would like to carry out a regression against more than one independent variable. Leyton (1956) studied the relation between growth and the mineral composition of the foliage of Japanese larch trees. A stand of trees had been planted in heterogeneous soil using variable planting stock. Leyton was concerned to discover the causes of the resulting variability in the height of the trees. Twenty-six trees were selected, ranging in height from 65 to 351 cm. From each tree one lateral shoot was taken from the uppermost whorl immediately below the terminal shoot and after drying the needles in an oven they were ground in a mill and analysed for nitrogen, phosphorus, potassium and residual ash. The data are given in Table 9.11.

We now want to do a regression of height against each of the other variables so that our model is

$$y_i = a + bN_i + cP_i + dK_i + eA_i + \varepsilon_i \qquad 9.33$$

and we can again find expressions for the coefficients of the fit that minimize the sums of squares of the residuals (Bliss, 1967, p. 308). The results are given in Table 9.12 where only the coefficients of nitrogen and potassium are significant. We might try omitting phosphorus and residual ash, but if we do this the other coefficients change, as in our polynomial regression, because the independent variables are not orthogonal. (To test if two variables, say N_i and K_i, are orthogonal, we calculate $\Sigma N_i K_i$ and if the answer is zero they are orthogonal. If we are able to choose the levels of the independent variables we choose them to ensure orthogonality. In the case of our larch trees, we simply have to use the measured values.)

Table 9.13 gives the correlation between each pair of variables. We see that the level of nitrogen is positively correlated with the level of each of the other variables. Any variable, in our case the height of the trees, that goes up and down with the level of nitrogen will also go up and down with the level of potassium, for example. In other words, some of the variation that is explained by variations in the level of nitrogen can also be explained by variations in the level of potassium. Consequently, if we first perform a regression against nitrogen alone and then perform a regression against nitrogen and potassium, some of the variation explained by nitrogen in the first regression will be explained by the potassium in the second regression and the coefficient for nitrogen will change.

We need to proceed carefully and there are two standard approaches to carrying out regressions when the independent variables are correlated. The first approach is **forward step-wise regression**. We carry out a regression against each independent

Table 9.11 The concentration, measured in parts million, of nitrogen, phosphorus, potassium and residual ash in the needles of 26 Japanese larch trees with their heights in centimetres

Nitrogen	Phosphorus	Potassium	Ash	Height
2.20	0.417	1.35	1.79	351
2.10	0.354	0.90	1.08	249
1.52	0.208	0.71	0.47	171
2.88	0.335	0.90	1.48	373
2.18	0.314	1.26	1.09	321
1.87	0.271	1.15	0.99	191
1.52	0.164	0.83	0.85	225
2.37	0.302	0.89	0.94	291
2.06	0.373	0.79	0.80	284
1.84	0.265	0.72	0.77	213
1.89	0.192	0.46	0.46	138
2.45	0.221	0.76	0.95	213
1.88	0.186	0.52	0.95	151
1.93	0.207	0.60	0.92	130
1.80	0.157	0.67	0.60	93
1.81	0.195	0.47	0.57	95
1.49	0.165	0.66	0.80	147
1.53	0.226	0.68	0.66	88
1.43	0.224	0.44	0.45	65
1.54	0.271	0.51	0.95	120
1.13	0.187	0.38	0.63	72
1.63	0.200	0.62	1.10	160
1.36	0.211	0.71	0.47	72
1.76	0.283	0.96	0.96	252
2.53	0.284	0.85	1.39	310
2.59	0.303	1.02	0.95	336

Table 9.12 The regression coefficients for the model given by Equation 9.33 applied to the data in Table 9.11

Chemical	N	P	K	A
Coefficient	97.76	256.97	126.57	40.28
s	24.57	169.91	46.43	36.61
t	3.98	1.51	2.73	1.10
*	***	ns	*	ns

variable separately and include the one that explains the greatest proportion of the variance. We then perform regression keeping this variable and including each of the others, one at a time, to see which variable explains the greatest proportion of the remaining variance. In other words, we find the most significant variable and include it. We then find the next most significant variable and include that. We proceed in this way until none of the excluded variables contributes significantly.

Table 9.13 Correlation coefficients for the concentrations of nitrogen, phosphorus, potassium and ash in Japanese larch trees

	Nitrogen	*Phosphorus*	*Potassium*	*Ash*
Nitrogen	1.00	0.60	0.55	0.65
Phosphorus	0.60	1.00	0.71	0.67
Potassium	0.55	0.70	1.00	0.67
Ash	0.65	0.67	0.67	1.00

The second approach is **backward step-wise regression**. We start by including all the variables and then drop the least significant variable. We then recalculate the contributions for the remaining variables and again drop the least significant, continuing in this way until all of the included variables are significant.

Unfortunately, there is not usually a single 'correct' answer and forward and backward step-wise regressions can leave us with different sets of variables. In practice it is worth doing both and if they give different answers you will have to judge between them on the basis of your knowledge and understanding of the biology of the problem.

Table 9.14 illustrates the forward step-wise regression. We start with none of the variables in the model and F-to-enter gives the significance level that we will obtain if we include each variable singly. Since we have 26 points, each of these F values should be compared with $F_{1,24}$ at the 5% level, which is 4.26, and each of them, taken on its own, is highly significant. Since nitrogen has the biggest F ratio, we include it first. The next part of the table shows the effect of each of the other variables, after allowing for the effect of nitrogen. Because nitrogen has already explained much of the variance, the contributions from each of the other variables is now much less

Table 9.14 F-to-enter, F-to-remove and the final model for the forward step-wise regression of the height of Japanese larch trees against the amount of nitrogen, potassium, phosphorus and residual ash. The table also gives the regression coefficients for the variables that are included in the regression

Variables in model	Coefficient	F-to-remove	Variables not in model	F-to-enter
			Nitrogen	48.6
			Phosphorus	35.9
			Potassium	40.3
			Ash	34.4
Nitrogen	183.4	48.6	Phosphorus	13.8
			Potassium	24.1
			Ash	9.4
Nitrogen	123.3	30.2	Phosphorus	3.39
Potassium	188.7	24.1	Ash	2.22

Table 9.15 F-to-enter, F-to-remove and the final model for the backward step-wise regression of the height of Japanese larch trees against the amount of nitrogen, potassium, phosphorus and residual ash

Variables in model	Coefficient	F-to-remove	Variables not in model	F-to-enter
Nitrogen	97.8	15.8		
Phosphorus	257.0	2.29		
Potassium	126.6	7.43		
Ash	40.3	1.21		
Nitrogen	107.8	22.1	Ash	1.21
Phosphorus	304.2	3.39		
Potassium	143.1	10.5		
Nitrogen	123.3	30.2	Phosphorus	3.39
Potassium	188.7	24.1	Ash	2.22

than before. We now compare each of the F values with $F_{1,23}$, since we have used up another degree of freedom, and this is 4.28. Potassium now makes the most significant contribution so we include it also. In the last part of the table we see that the two remaining variables are no longer significant since the variance that they would have explained is already accounted for by nitrogen and potassium, and we omit them from the model. We also see in Table 9.14 that when nitrogen is included on its own, the regression coefficient is 183.4, but when we include potassium in the regression, the coefficient for nitrogen is reduced to 123.3 because part of the increase due to nitrogen is now explained by the increase due to potassium.

Table 9.15 shows the corresponding sequence of steps for the backward stepwise regression and, in this case, we end by selecting the same two variables. Table 9.16 is the ANOVA for the final model.

Leyton (1956) concludes from his analysis of these data that the trees are suffering from a lack of nitrogen and potassium so that in those areas where the concentrations of these nutrients happen to be relatively high, the trees grow significantly taller. This experiment illustrates the care that must be taken when drawing conclusions from regressions. If Leyton had measured phosphorus but not nitrogen or potassium, it would have appeared that a significant part of the variation in growth could be explained on the basis of a phosphorus deficiency whereas the truth seems to be that the trees are deficient in nitrogen and potassium and that the concentration of phosphorus is correlated with both of these.

9.8 SUMMARY

ANOVA and regression are closely related and we can formulate ANOVA problems as regression problems or analyse regression problems using ANOVA. We use ANOVA when the independent variables are categorical. We use regression when the independent

Table 9.16 Parameters for the regression of the height of Japanese larch trees against the concentration of nitrogen and potassium

Parameter	Value	s	t	P	*
Intercept	− 181	36.9	− 4.9	0.0001	***
Nitrogen	123	22.4	5.5	0.0000	***
Potassium	189	38.4	4.9	0.0001	***

Source	Sum of squares	Degrees of freedom	Mean square	F-ratio	Significance level, P	*
Model	191 184	2	95592	59.8	0.0000	***
Error	36 771	23	1599			
Total	227 954	25		$R^2 = 0.84$		

variables are measured on a quantitative scale. Sometimes the independent variables may be of both kinds, some qualitative, some quantitative. In this case we can carry out a combined analysis called analysis of covariance or ANCOVA, in which the covariates are the quantitative variables. Brownbridge's armyworm data are an example that we might choose to analyse using ANCOVA with the treatments as categorical variables and time as a continuous variable against which we carry out a regression (see Example 9.10.6). In fact, we can go further and treat ANOVA, regression and the combination, ANCOVA, as particular cases of a wider approach called **generalized linear modelling** (Dobson, 1983; McCullogh and Nelder, 1983). It is important, however, to be clear as to just what each technique does and it is useful to think of ANOVA and regression as separate but related techniques.

The concept of interactions applies to regression just as to ANOVA. In regression problems we simply include new variables that are products of the original variables. In the example of Japanese larch trees we might, for example, have tried a term of the form $N \times K$ to see if the response to the levels of nitrogen depends on the levels of potassium and vice versa.

An important extension of these ideas involves the use of weights in ANOVA and regression. We often have data for which the different points are measured with differing accuracy. In this case, instead of finding the model that minimizes the sum of the squares of the residuals, $\Sigma \varepsilon_i^2$, we find the model that minimizes $\Sigma \varepsilon_i^2 w_i$ where we choose the weights w_i so that the more precise observations are given greater weight than the less precise observations. The best choice, in a least-squares sense, is to set the weights equal to the reciprocal of the variance so that if you know how the variance varies across your data you can weight the separate measurements accordingly (Hays, 1988, p. 624).

The ideas discussed in this book contain many of the most important principles of statistical analysis. However, there are a number of techniques collectively known as multivariate methods. **Multivariate analysis of variance**, **MANOVA** deals with problems in which we have more than one response variable. **Cluster analysis** allows us to group individuals, for each of which a number of variables have been measured, into hierarchical clusters starting from those that are most similar. **Discriminant analysis** allows us to find the combination of a set of variables that will best discriminate between the individuals of interest. **Principle components analysis** provides a method for combining a large number of variables into a smaller number of variables, each of which is a linear combination of the original variables. For further reading, the book by Jeffers (1978) provides a good introduction and the books by Manly (1986) and Green (1978) tell you how to carry out multivariate analyses.

The best advice I can leave you with comes from Max Born (Margenau, 1950): 'I believe that there is no philosophical high-road in science with epistemological sign-posts. No, we are in a jungle and find our way by trial and error, building our road behind us as we proceed. We do not find sign-posts at cross-roads, but our own scouts erect them to help the rest ... My advice to those who wish to learn the art of scientific prophecy is not to rely on abstract reason, but to decipher the language of Nature from Nature's documents, the facts of experience.'

9.9 APPENDIX

9.9.1 Regression sums of squares

We wish to partition the sums of squares of the points in a linear regression into a contribution from the fitted line and an error term. Since we are concerned with the variation about the mean, as in the standard ANOVA, we subtract the mean value of the dependent variable from each measured value y_i. Similarly, we can subtract the mean value of the independent variable, from each measured value of x_i. Then Σy_i and Σx_i are both zero and from Equation 9.9 a is also zero. The equation of the regression line is then

$$\hat{y}_i = bx_i \qquad\qquad 9.34$$

and the middle term on the right-hand side of Equation 9.13 is

$$\Sigma \hat{y}_i \varepsilon_i = \Sigma bx_i(bx_i - y_i) = b\left(b\Sigma x_i^2 - \Sigma x_i y_i\right). \qquad 9.35$$

However, from Equation 9.10 with Σy_i and Σx_i equal to zero

$$b = \Sigma x_i y_i / \Sigma x_i^2 \qquad\qquad 9.36$$

so that the right-hand side of Equation 9.35, and hence $\Sigma \hat{y}_i \varepsilon_i$ and the middle term on the right-hand side of Equation 9.13 are zero.

9.9.2 Variance of the slope

To calculate the variance of the slope in a linear regression we will again subtract the mean values of x and y from each data point since this will not affect the slope. Then

$$b = \Sigma xy / \Sigma x^2 \qquad\qquad 9.37$$

as in Appendix, section 9.9.1. But

$$V(ay) = a^2 V(y) \qquad 9.38$$

so that

$$V(b) = \Sigma x^2 V(y)/(\Sigma x^2)^2 = V(y)/\Sigma x^2 = s_p^2/\Sigma x^2 \qquad 9.39$$

as in Equation 9.21 with x equal to $x_i - \bar{x}$.

9.9.3 Variance of the fitted line

To calculate the variance of the fitted line we again subtract the mean value of x and y from each data point. Then

$$(y_i - \bar{y}) = b(x_i - \bar{x}) \qquad 9.40$$

since a is zero. Then

$$V(y_i) = V(\bar{y}_i) + V(b)(x_i - \bar{x})^2 \qquad 9.41$$

since we assume that the only errors are in y and $V(x) = 0$. Therefore

$$V(y_i) = s_m^2 + s_b^2(x_i - \bar{x})^2 \qquad 9.42$$

9.10 EXERCISES

1. Use Equations 9.19 to 9.22 to calculate 95% confidence limits for the residuals plotted in Fig. 9.7b obtained by fitting Equation 9.25 to the data given in Table 9.5. Decide if the large residual on day 9 differs significantly from the fitted line. (Since you are only considering the largest residual you should use a significance level of 5/14% to obtain a 95% confidence band.)

2. In studies of evolution it is important to be able to fix the rate at which evolutionary changes occur. Given two related species, we should then be able to estimate the time since they began to diverge. Fitch counted the mean number of nucleotide substitutions in seven proteins in 15 related species for which it was possible to determine the time at which they diverged from the fossil record, with the results given in Table 9.17 (Ruse, 1982).

(a) Calculate the mean number of nucleotide substitutions per million years and determine the time at which two species differing by an average of 40 nucleotide substitutions diverged. (b) Calculate 95% confidence limits for the points and place error limits on the answer to (a). [Think carefully about which variable to take as the independent variable.]

3. Large islands generally support more species than comparable small islands and the number of species, N, generally follows a power-law relationship with the area of the habitat, A, so that

$$N = aA^z. \qquad 9.43$$

Table 9.17 The mean number of nucleotide substitutions in seven proteins in 15 related species and the time since they diverged

Nucleotide substitutions	Time/M years	Nucleotide substitutions	Time/M years
7.91	1.65	40.29	53.14
4.46	7.49	48.92	63.62
4.17	11.23	46.76	61.23
1.44	13.77	56.83	75.60
17.27	15.87	66.19	70.36
7.19	19.76	73.67	90.57
15.83	32.93	97.84	121.26
33.09	41.92		

The exponent in the power law, z, typically lies between 0.24 and 0.33 (Pianka, 1983). Davis *et al.* (1988) studied assemblages of forest-dwelling species of small, non-flying mammals on the mountains of New Mexico and adjacent parts of Arizona. These montane islands of forest are surrounded by woodlands and grasslands that isolate one from another. Their results are given in Table 9.18. Carry out a regression of *ln N* against *ln A*, decide if the data show a power-law relationship and determine the value of z. [Hint: Since the logarithm of 0 is $-\infty$ there is a problem when there are no species. To avoid this Davis *et al.* (1988) added 1 to each of the species numbers before taking logarithms.]

The authors argue that species richness decreases with the distance from the original source of the various species, which they assume to be in the southern extension of the Rocky Mountains. They therefore also measured the shortest distance from there to the individual forests while remaining within suitable habitats and this is given as the 'isolation' in Table 9.18. Carry out a regression of the residuals from the previous

Table 9.18 The area, the number of species and the isolation of 12 montane forest islands

Forest	Area/km^2	No. species	Isolation/km
1	21 813	10	241
2	2 872	7	217
3	346	6	75
4	2 575	6	292
5	885	4	72
6	2 090	4	378
7	1 279	3	173
8	750	3	281
9	173	3	271
10	205	2	269
11	57	0	475
12	8	0	392

Table 9.19 The mean antler length and shoulder height of various cervine deer. All measurements in inches. The last two numbers, shown in bold, are the values for the Irish Elk

Antler	63.6	51.1	48.7	44.9	42.4	39.5
Shoulder	65.0	50.4	55.0	54.5	45.5	50.9
Antler	40.2	36.1	34.0	31.8	28.3	26.3
Shoulder	36.8	40.9	45.7	36.3	34.8	32.5
Antler	22.9	23.8	20.9	13.4	9.7	**94.4**
Shoulder	39.1	27.5	34.0	27.1	27.0	**72.9**

fit against the isolation and see if you agree that the isolation also influences species richness. (For discussions of the species–area relationship see Diamond and May, 1981; MacArthur and Wilson, 1963; Brown and Lomolino, 1989.)

4. The Irish Elk, which Gould (1974, 1977) points out is neither exclusively Irish nor an elk, became extinct in Ireland 11 000 years ago. It has become famous for the prodigious size of its antlers whose span could reach 12 feet. Early in this century the Irish Elk was the centre of a fierce dispute between proponents and opponents of Darwin's theory of evolution for it was hard to see how such large antlers could confer a selective advantage on the animal and seemed more likely to lead to the animals' extinction. To decide if the antlers really were too big in proportion to the body size, Gould (1973) measured the spread of the antlers and the shoulder heights of various cervine deer, the group to which the Elk belongs, with the results given in Table 9.19. Plot a graph of the logarithms of the spread of the antlers against the logarithms of the shoulder heights. Fit a straight line to all of the data excluding that of the Irish Elk and decide if you agree with Gould that the antlers of the Irish Elk are only as big as one would expect in a deer of that size.

Calculate 95% confidence limits for the points and determine how big (or small) the average spread of the antlers would have to be in order to reject the hypothesis that the antlers are about as big as one would expect. Do the data provide support for Gould's argument?

5. MacFadden (1986) carried out a study to investigate the rate at which the body size of horses has changed over evolutionary time. It is difficult to estimate the body size of fossil horses as only fragments are usually found. MacFadden therefore began by studying the relationship between various bodily dimensions for New World Equidae in order to find the most useful measure of body size one could obtain from fossil remains. In Table 9.20, for example, the head–body length is compared with the length of the row of cheek-teeth. In general, the cheek-teeth are among the better preserved parts of fossil skeletons. The data have been read off a graph in MacFadden's paper. Carry out a regression of the head–body length against the length of the row of cheek-teeth and see if the latter can be used as a good estimator of the former.

Using the data in this way MacFadden was able to estimate the body sizes of 40 fossil horses covering about 55 million years as shown in Table 9.21. Plot a graph of the logarithm of the body mass against the age of the fossils. MacFadden concluded

Table 9.20 HBL is the head–body length measured from the tip of the nose to the beginning of the first caudal vertebra and TRL is the overall length of the row of cheek-teeth from the antero-most occlusal enamel of tooth P2 to postero-most occlusal enamel of tooth M3. All measurements in millimetres

TRL	39.4	43.6	45.6	46.7	68.9	67.2	66.9
HBL	648	648	691	648	843	911	988
TRL	84.7	85.0	90.0	105	115	148	140
HBL	1221	1306	1349	1306	1603	1709	1752
TRL	140	158	161	173	170	173	170
HBL	1900	1900	2109	2240	2304	2304	2346
TRL	183	183	183	188	189	193	190
HBL	2648	2588	2542	2499	2448	2448	2448

Table 9.21 The mean body masses and the ages of 40 fossil horses

Number	1	2	3	4	5
Myr	57.53	52.77	52.77	50.26	50.26
Mass/kg	31.87	25.75	40.16	32.74	23.57
Number	6	7	8	9	10
Myr	45.49	45.53	31.91	30.64	25.87
Mass/kg	31.87	44.52	47.58	48.02	53.25
Number	11	12	13	14	15
Myr	23.40	21.91	20.85	18.43	18.38
Mass/kg	101.71	115.24	93.41	78.13	47.58
Number	16	17	18	19	20
Myr	18.38	11.06	14.81	13.49	15.96
Mass/kg	132.70	412.94	199.92	268.45	69.84
Number	21	22	23	24	25
Myr	15.79	15.83	13.45	11.06	13.62
Mass/kg	85.56	53.69	78.57	122.66	139.68
Number	26	27	28	29	30
Myr	8.55	14.77	11.06	13.57	13.49
Mass/kg	61.11	108.69	153.21	170.24	161.51
Number	31	32	33	34	35
Myr	12.34	8.72	6.04	6.04	3.62
Mass/kg	137.50	235.71	244.01	313.85	84.68
Number	36	37	38	39	40
Myr	5.15	2.38	2.47	2.47	2.47
Mass/kg	434.76	436.94	501.98	351.83	457.02

Table 9.22 Numbers of armyworm larvae sampled in each plot on each day after the start of treatment. Note that treatment C actually started one day after treatments A and B. ● indicates missing data

Row		1	1	1	2	2	2	3	3	3
Column		1	2	3	1	2	3	1	2	3
Treatment		B	A	C	C	B	A	A	C	B
	0	331	425	420	287	390	290	267	214	197
No.	1	242	348	200	64	223	173	138	63	178
on	2	213	306	43	26	109	115	105	31	71
day	3	●	●	20	8	●	●	●	10	●
	4	154	204	●	●	102	106	48	●	88

that there may not have been much evolutionary change up to 35 million years ago but that since then the body size has been increasing at a rate of 5.6%/Myr if we exclude species 26 and 35, which seem to be especially small. These two species do in fact belong to a different *Nannipus* species from the others so that it might be reasonable to exclude them. Carry out a regression analysis to see if you agree with his conclusions.

6. The actual data that Brownbridge (1988) collected in his experiment on the biocidal effects of *Bacillus thuringiensis* on the armyworm *Spodoptera exempta* are given in Table 9.22. The experiment used a Latin square design with three rows and three columns laid out in the field under study. Considerable attention was given to the sampling process. Within each plot 20 quadrats, each 20cm × 20cm, were chosen at random on each day and the numbers of larvae in the quadrats were counted. The outer 2 m of each plot were excluded from sampling to minimize effects arising from movement of larvae into or out of each plot. As a control, 30 quadrats were chosen at random on each day from the untreated areas around and between the treated plots (Table 1.3).

Unfortunately, as a result of heavy rainfall at the time of spraying, only two treatments (A and B) were applied on Day 0. Treatment C was applied for the first time on Day 1. Furthermore, counting had to be abandoned early on Day 3 because the farm manager wished to spray the experimental area with pyrethrum as there were still a lot of armyworms in the untreated areas and the risk of damage to the rest of his crop was too great. On that day counts were made only for the three plots under treatment C. After summing the larval numbers over the quadrats in each experimental plot, the data given in Table 9.23 were recorded.

The data were analysed, after taking logarithms, using *MiniTab*, which is able to handle missing data points. The rows, columns and treatments were qualitative factors and time was specified as the covariate so that this is an example of ANCOVA. Table 9.23 gives the result of the analysis.

Because of the missing values, the factors in the analysis are not orthogonal. In the means table, 'Time' gives the average slope and 'Time × Tr.X' gives the deviation from the average slope for treatment X.

Table 9.23 Analysis of Brownbridge's armyworm data using the general linear modelling facility in MiniTab.

Source	Sums of squares	Degrees of freedom	Mean square	F ratio	Significance level, P	*
Rows	3.788	2	1.894	34.69	< 0.001	***
Columns	0.375	2	0.188	3.44	0.047	*
Treatments	0.031	2	0.015	0.28	0.755	ns
Time	17.905	1	17.905	327.94	< 0.001	***
Time × Treat.	7.683	2	3.842	70.36	< 0.001	***
Error	1.420	26	0.055			
	Coefficient	s	t	P	*	
Constant	5.630	0.062	90.82	< 0.001	***	
Time	− 0.533	0.029	− 18.11	< 0.001	***	
Time × Tr.A	0.255	0.040	6.46	< 0.001	***	
Time × Tr.B	0.285	0.040	7.22	< 0.001	***	
Time × Tr.C	− 0.541	0.040	13.53	< 0.001	***	

Make sure that you understand how each term in the table is calculated. Determine the slopes under each treatment and their confidence limits and compare these with the values given in Table 9.4. (When I presented the data in Table 9.1 I used the fit from Table 9.23 to determine best estimates of the missing values and then added suitably chosen random numbers to these estimates to avoid the problem of missing values.) Note that the sums of squares and mean squares given here are given as the adjusted sums of squares (ASS) and adjusted mean squares (AMS) in MiniTab.

10

Tables

These tables were calculated using routines in the *NAG* Library provided by *NAG* Ltd. (For address see Preface.)

Table 10.1 Critical values for Student's *t* distribution. *P* gives the value of the c.d.f., that is the probability that a number chosen from a *t* distribution is less than the value in the body of the table for the appropriate number of degrees of freedom. At the foot of the table *SL 1* and *2* give the significance levels for one- and two-tailed tests, respectively. Student's *t* distribution with an infinite number of degrees of freedom is identical to the standard normal distribution

Degrees of			*P*			
freedom	*0.95*	*0.975*	*0.99*	*0.995*	*0.999*	*0.9995*
1	6.31	12.71	31.82	63.66	318.31	636.59
2	2.92	4.30	6.96	9.92	22.33	31.60
3	2.35	3.18	4.54	5.84	10.21	12.92
4	2.13	2.78	3.75	4.60	7.17	8.61
5	2.02	2.57	3.36	4.03	5.89	6.87
6	1.94	2.45	3.14	3.71	5.21	5.96
7	1.89	2.36	3.00	3.50	4.79	5.41
8	1.86	2.31	2.90	3.36	4.50	5.04
9	1.83	2.26	2.82	3.25	4.30	4.78
10	1.81	2.23	2.76	3.17	4.14	4.59
15	1.75	2.13	2.60	2.95	3.73	4.07
20	1.73	2.09	2.53	2.85	3.55	3.85
30	1.70	2.04	2.46	2.75	3.39	3.65
60	1.67	2.00	2.39	2.66	3.23	3.46
120	1.66	1.98	2.36	2.62	3.16	3.37
Infinity	1.65	1.96	2.33	2.58	3.09	3.29
SL 1 tail	*5%*	*2.5%*	*1.0%*	*0.5%*	*0.1%*	*0.05%*
SL 2 tail	*10%*	*5.0%*	*2.0%*	*1.0%*	*0.2%*	*0.10%*

Tables

Table 10.2 Critical values for the χ^2 distribution. P gives the value of the c.d.f., that is the probability that a number chosen from a χ^2 distribution is less than the value in the body of the table for the appropriate number of degrees of freedom. SL gives the significance level for a one-tail test. To evaluate significance levels for a critical value of χ^2 equal to x, say, when the number of degrees of freedom exceeds 100, calculate $y = (2x)^{1/2} - (2f - 1)^{1/2}$ and then look up the significance level of y for a standard normal distribution

Degrees of freedom			P			
	0.05	0.1	0.9	0.95	0.99	0.999
1	0.0039	0.016	2.71	3.84	6.63	10.83
2	0.103	0.211	4.61	5.99	9.21	13.82
3	0.352	0.584	6.25	7.81	11.34	16.27
4	0.711	1.06	7.78	9.49	13.28	18.47
5	1.15	1.61	9.24	11.07	15.09	20.52
6	1.64	2.20	10.64	12.59	16.81	22.46
7	2.17	2.83	12.02	14.07	18.48	24.32
8	2.73	3.49	13.36	15.51	20.09	26.12
9	3.33	4.17	14.68	16.92	21.67	27.88
10	3.94	4.87	15.99	18.31	23.21	29.59
11	4.57	5.58	17.28	19.68	24.72	31.26
12	5.23	6.30	18.55	21.03	26.22	32.91
13	5.89	7.04	19.81	22.36	27.69	34.53
14	6.57	7.79	21.06	23.68	29.14	36.12
15	7.26	8.55	22.31	25.00	30.58	37.70
16	7.96	9.31	23.54	26.30	32.00	39.25
17	8.67	10.09	24.77	27.59	33.41	40.79
18	9.39	10.86	25.99	28.87	34.81	42.31
19	10.12	11.65	27.20	30.14	36.19	43.82
20	10.85	12.44	28.41	31.41	37.57	45.31
21	11.59	13.24	29.62	32.67	38.93	46.80
22	12.34	14.04	30.81	33.92	40.29	48.27
23	13.09	14.85	32.01	35.17	41.64	49.73
24	13.85	15.66	33.20	36.42	42.98	51.18
25	14.61	16.47	34.38	37.65	44.31	52.62
26	15.38	17.29	35.56	38.89	45.64	54.05
28	16.93	18.94	37.92	41.34	48.28	56.89
30	18.49	20.60	40.26	43.77	50.89	59.70
40	26.51	29.05	51.81	55.76	63.69	73.40
50	34.76	37.69	63.17	67.50	76.15	86.66
60	43.19	46.46	74.40	79.08	88.38	99.61
70	51.74	55.33	85.53	90.53	100.43	112.32
80	60.39	64.28	96.58	101.88	112.33	124.84
90	69.13	73.29	107.57	113.14	124.12	137.21
100	77.93	82.36	118.50	124.34	135.81	149.45
SL	95%	90%	10%	5%	1%	0.1%

Table 10.3 Critical values for the F distribution with f_1 and f_2 degrees of freedom, for a cumulative probability of 0.95 or a significance level of 5%

f_2							f_1						
	1	2	3	4	5	6	8	12	15	20	30	60	Infinity
1	161.45	199.50	215.82	224.75	230.36	234.21	239.13	244.18	246.24	248.31	250.41	252.52	254.31
2	18.51	19.00	19.16	19.25	19.30	19.33	19.37	19.41	19.43	19.45	19.46	19.48	19.50
3	10.13	9.55	9.28	9.12	9.01	8.94	8.85	8.74	8.70	8.66	8.62	8.57	8.53
4	7.71	6.94	6.59	6.39	6.26	6.16	6.04	5.91	5.86	5.80	5.75	5.69	5.63
5	6.61	5.79	5.41	5.19	5.05	4.95	4.82	4.68	4.62	4.56	4.50	4.43	4.36
6	5.99	5.14	4.76	4.53	4.39	4.28	4.15	4.00	3.94	3.87	3.81	3.74	3.67
7	5.59	4.74	4.35	4.12	3.97	3.87	3.73	3.57	3.51	3.44	3.38	3.30	3.23
8	5.32	4.46	4.07	3.84	3.69	3.58	3.44	3.28	3.22	3.15	3.08	3.00	2.93
9	5.12	4.26	3.86	3.63	3.48	3.37	3.23	3.07	3.01	2.94	2.86	2.79	2.71
10	4.96	4.10	3.71	3.48	3.33	3.22	3.07	2.91	2.85	2.77	2.70	2.62	2.54
12	4.75	3.89	3.49	3.26	3.11	3.00	2.85	2.69	2.62	2.54	2.47	2.38	2.30
14	4.60	3.74	3.34	3.11	2.96	2.85	2.70	2.53	2.46	2.39	2.31	2.22	2.13
16	4.49	3.63	3.24	3.01	2.85	2.74	2.59	2.42	2.35	2.28	2.19	2.11	2.01
20	4.35	3.49	3.10	2.87	2.71	2.60	2.45	2.28	2.20	2.12	2.04	1.95	1.84
25	4.24	3.39	2.99	2.76	2.60	2.49	2.34	2.16	2.09	2.01	1.92	1.82	1.71
30	4.17	3.32	2.92	2.69	2.53	2.42	2.27	2.09	2.01	1.93	1.84	1.74	1.62
40	4.08	3.23	2.84	2.61	2.45	2.34	2.18	2.00	1.92	1.84	1.74	1.64	1.51
60	4.00	3.15	2.76	2.53	2.37	2.25	2.10	1.92	1.84	1.75	1.65	1.53	1.39
120	3.92	3.07	2.68	2.45	2.29	2.18	2.02	1.83	1.75	1.66	1.55	1.43	1.25
Infinity	3.84	3.00	2.60	2.37	2.21	2.10	1.94	1.75	1.67	1.57	1.46	1.32	1.00

Table 10.4 Critical values for the F distribution with f_1 and f_2 degrees of freedom for points on the c.d.f. corresponding to a cumulative probability of 0.99 or a significance level of 1%

f_2	f_1 1	2	3	4	5	6	8	12	15	20	30	60	Infinity
1	4051.86	4999.51	5403.47	5624.76	5763.86	5859.22	5981.33	6106.61	6157.59	6209.04	6260.97	6313.36	6365.88
2	98.50	99.00	99.17	99.25	99.30	99.33	99.75	99.83	99.87	99.90	99.93	99.97	99.50
3	34.11	30.82	29.46	28.71	28.24	27.91	27.49	27.05	26.87	26.69	26.50	26.32	26.13
4	21.20	18.00	16.69	15.98	15.52	15.21	14.80	14.37	14.20	14.02	13.84	13.65	13.46
5	16.26	13.27	12.06	11.39	10.97	10.67	10.29	9.89	9.72	9.55	9.38	9.20	9.02
6	13.74	10.92	9.78	9.15	8.75	8.47	8.10	7.72	7.56	7.40	7.23	7.06	6.88
7	12.25	9.55	8.45	7.85	7.46	7.19	6.84	6.47	6.31	6.16	5.99	5.82	5.65
8	11.26	8.65	7.59	7.01	6.63	6.37	6.03	5.67	5.52	5.36	5.20	5.03	4.86
9	10.56	8.02	6.99	6.42	6.06	5.80	5.47	5.11	4.96	4.81	4.65	4.48	4.31
10	10.04	7.56	6.55	5.99	5.64	5.39	5.06	4.71	4.56	4.41	4.25	4.08	3.91
12	9.33	6.93	5.95	5.41	5.06	4.82	4.50	4.16	4.01	3.86	3.70	3.54	3.36
14	8.86	6.51	5.56	5.04	4.69	4.43	4.14	3.80	3.66	3.51	3.35	3.18	3.00
16	8.53	6.23	5.29	4.77	4.44	4.20	3.89	3.55	3.41	3.26	3.10	2.93	2.75
20	8.10	5.85	4.94	4.43	4.10	3.87	3.56	3.23	3.09	2.94	2.78	2.61	2.42
25	7.77	5.57	4.68	4.18	3.85	3.63	3.32	2.99	2.85	2.70	2.54	2.36	2.17
30	7.56	5.39	4.51	4.02	3.70	3.47	3.17	2.84	2.70	2.55	2.39	2.21	2.01
40	7.31	5.18	4.31	3.83	3.51	3.29	2.99	2.66	2.52	2.37	2.20	2.02	1.80
60	7.08	4.98	4.13	3.65	3.34	3.12	2.82	2.50	2.35	2.20	2.03	1.84	1.60
120	6.85	4.79	3.95	3.48	3.17	2.96	2.66	2.34	2.19	2.03	1.86	1.66	1.38
Infinity	6.63	4.61	3.78	3.32	3.02	2.80	2.51	2.18	2.04	1.88	1.70	1.47	1.00

References

Ashby, E. and Oxley, T. A. (1935) 'The interaction of factors in the growth of Lemna VI: an analysis of the influence of light intensity and temperature on the assimilation rate and the rate of frond multiplication', *Annals of Botany*, **49**, 309–36.

Ayala, F. J. and Kiger, J. A. Jr. (1980) *Modern Genetics*, Benjamin/Cummings, Menlo Park, California.

Banic, S. (1975) *Nature*, **258**, 153–4. Discussed by Wardlaw, A. C. (1985) *Practical Statistics for Experimental Biologists*, John Wiley, New York, p. 103.

Bateson, W. (1909) *Mendel's Principles of Heredity*, Cambridge University Press, Cambridge.

Bliss, C. I. (1967) *Statistics in Biology*, McGraw-Hill, New York.

Boslough, J. (1984) *Beyond the Black Hole: Stephen Hawking's Universe*, Fontana, London, p. 38.

Brightwell, R., Dransfield, R. D. and Kyorku, C. (1991) 'Development of a low cost tsetse trap and odour baits for *Glossina pallidipes* and *G. longipennis*', *Medical and Veterinary Entomology*, **5** 153–64.

Broad, W. and Wade, N. (1985) *Betrayers of the Truth: Fraud and Deceit in Science*, Oxford University Press, Oxford, p. 31.

Brown, J. H. and Lomolino, M. V. (1989) 'Independent discovery of the equilibrium theory of island biogeography', *Ecology*, **70**, 1954–7.

Brown, P. (1991) 'Canada to test Kenya's disputed HIV drug', *New Scientist*, **12**, January, p. 27.

Brownbridge, M. (1988) 'Field trials of *Bacillus thurungiensis* against the African armyworm *Spodoptera exempta* June 8th–June 12th 1988', Internal Report, International Centre for Insect Physiology and Ecology, P.O. Box 30772, Nairobi, Kenya.

Bulmer, M. G. (1979) *Principles of Statistics*, Dover, New York.

Burkhardt, R. W. (1981) 'Species', in Bynum, F., Browne, E. J. and Porter, R. (eds) (1981) *Dictionary of the History of Science*, Macmillan Press, London, p. 396. See also, Ronan, C. A. (1983) *The Cambridge Illustrated History of the World's Science*, Cambridge University Press, Cambridge, p. 398.

Cain, A. J. and Sheppard, P. M. (1952) 'The effects of natural selection on body colour in the land snail *Cepaea nemoralis*', *Heredity*, **6**, 217.

Cain, A. J. and Sheppard, P. M. (1954) 'Natural selection in *Cepaea nemoralis*', *Genetics*, **39**, 89–116.

Challier, A., Eyraud, M., Lafaye, A. and Laveissière, C. (1977) 'Amélioration du rendement du piege biconique pour glossines (Diptera, Glossinidae) par l'emploi d'un cone inférieur blue', *Cahiers ORSTOM, série Entomologie médicale et Parasitologie*, **15**, 283–6.

Cronin, H. (1991) *The Ant and the Peacock: Altruism and Sexual Selection from Darwin to Today*, Cambridge University Press, Cambridge.

Crooke, W. M. and Knight, A. H. (1965) 'The relationship between nickel-toxicity symptoms and the absorption of iron and nickel', *Annals of Applied Biology*, **43**, 454–64. Discussed by Bliss, C. I. (1967) in *Statistics in Biology*, Volume 1, McGraw-Hill, New York, p. 37.

Cushny, A. R. and Peebles, A. R. (1905) 'The action of optical isomers II. Hyoscines', *Journal of Physiology*, **32**, 501–10.

Darwin, C. (1906) *Origin of Species*, John Murray, London.

Darwin, C. (1958) *Autobiography*, edited by N. Barlow, Collins, London, p. 120, cited by A. Flew in the introduction to *An Essay on the Principle of Population*, Penguin, Harmondsworth, p. 49.

Davis, R. Dunford, C. and Lomolino, M. V. (1988) 'Montane mammals of the American southwest: The possible influence of post-Pleistocene colonization', *Journal of Biogeography*, **15**, 841–8.

Diamond, J. M. and May, R. M. (1981) 'Island biogeography and the design of natural reserves', in May, R. M. (Ed.) (1981) *Theoretical Ecology*, Blackwell Scientific, Oxford, p. 228–52.

Dobson, A. J. (1983) *An Introduction to Statistical Modelling*, Chapman & Hall, London.

Dobzhansky, T. and Pavlovsky, O. (1957) 'An experimental study of the interaction between genetic drift and natural selection', *Evolution*, **11**, 311–19.

Dransfield, R. D., Brightwell, R., Kyorku, C. and Williams, B. G. (1990) 'Control of tsetse populations using traps at Nguruman, southwest Kenya', *Bulletin of Entomological Research*, **80**, 265–76.

Dransfield, R. D., Brightwell, R. and Williams, B. G. (1991) 'Control of tsetse flies and trypano-somiasis—Myth or Reality?', *Parasitology Today*, **7**, 287–91.

Einstein, A. (1950) 'The fundaments of theoretical physics', *Out of My Later Years*, Philosophical Library, New York, p. 98.

Einstein, A. (1954) 'Geometry and experience', in *Ideas and Opinions*, Crown, New York, p. 233.

Einstein, A. (1978) *The Meaning of Relativity*, Chapman & Hall, London, p. 1.

Farrow, R. A. (1979) 'Population dynamics of the Australian plague locust *Chortoicetes terminifera* in central western New South Wales. I. Reproduction and migration in relation to weather', *Australian Journal of Zoology*, **27**, 717–45.

Feynman, R. (1980) *The Character of a Physical Law*, MIT Press, Cambridge, Massachusetts, p. 34.

Fisher, R. A. (1936) 'Has Mendel's work been rediscovered?', reprinted from *Annals of Science*, **1**, 115–37 and in Bennet, J. H. (Ed.) (1965) *Experiments in Plant Hybridization*, Oliver and Boyd, Edinburgh.

Fisher, R. A. (1965) 'Marginal comments on Mendel's paper', in Bennet, J. H. (Ed.) (1965) *G. Mendel: Experiments in Plant Hybridization*, Oliver and Boyd, Edinburgh.

Freund, J. E. (1972) *Mathematical Statistics*, Prentice-Hall, London.

Galton, F. (1889) *Natural Inheritance*, Macmillan, London. Cited by Snedecor, G. W. and Cochran, W. G. (1987) in *Statistical Methods*, Iowa State University Press, Ames, p. 171.

Gardiner, M. (1970) 'The fantastic combinations of John Conway's new solitaire game of life', *Scientific American*, **223**, 120–3.

Gardiner, M. (1971) 'On cellular automata, self-reproduction, the Garden of Eden and the game of Life', *Scientific American*, **224** 112–17.

Gardiner, M. (1983) *Wheels, Life and other Mathematical Amusements*, Freeman, New York.

Giessler, A. (1889) 'Beiträge zur Frage des Geslechtsverhältnisses der Geboren', *Z. K. Sächsischen Statistischen Bureaus*, **35**, 1. Discussed by M. G. Bulmer (1979) in *Principles of Statistics*, Dover, New York, p. 88.

Gould, S. J. (1973) 'Positive allometry of antlers in the Irish Elk *Megaloceros giganteus*', *Nature*, **244**, 375–6. Discussed by Ruse, M. (1982) in *Darwinism Defended*, Addison-Wesley, New York, p. 149.

Gould, S. J. (1974) 'The origin and function of "bizarre" structures: antlers and skull size in the "Irish elk" *Megaloceros giganteus*', *Evolution* **28**, 191–220.

Gould, S. J. (1977) 'The misnamed, mistreated and misunderstood Irish Elk', in *Ever Since Darwin*, Norton, New York, p. 79, 85.

Gould, S. J. (1980) *The Panda's Thumb*, Norton, New York, p. 225.

Gould, S. J. (1981) *The Mismeasure of Man*, Pelican Books, London.

Gould, S. J. (1986) 'Evolution as fact and theory', *Hen's Teeth and Horse's Toes*, Penguin Books, Harmondsworth, England, p. 255.

Green, P. E. (1978) *Analysing Multivariate Data*, Dryden Press, Hinsdale, Illinois.

Greenwood, M. and Yule, G. U. (1920) 'An inquiry into the nature of frequency distributions of multiple happenings', *Journal of the Royal Statistical Society*, **83**, 255–79. Discussed by Sokal, R. R. and Rohlf, F. J. (1981) *Biometry*, Freeman, New York, p. 94.

Hacking, I. (1979) *Logic of Statistical Inference*, Cambridge University Press, Cambridge.

Hays, W. L. (1988) *Statistics*, Holt, Rinehart and Winston, Orlando.

Healy, M. J. R. (1981) 'Choice of sample size in parasitological experiments', *Parasitology Today*, **3**, 91–4.

Hunter, D. M. and Cosenzo, E. L. (1990) 'The origin of plagues and recent outbreaks of the South American locust, *Schistocerca cancellata* (Orthoptera: Acrididae) in Argentina', *Bulletin of Entomological Research*, **80**, 295–300.

Jammer, M. (1974) *The Philosophical Foundations of Quantum Mechanics*, John Wiley, New York, p. 155. Letter from Albert Einstein to Max Born, 4 December 1926.

Jeffers, J. N. R. (1978) *An Introduction to Systems Analysis: With Ecological Applications*, Edward Arnold, London.

Kendall, M. and Stuart, A. (1983) *The Advanced Theory of Statistics*, Volume II, Charles Griffin, London, pp. 110 and 134.

Kendall, M., Stuart, A. and Ord, J. K. (1983) *The Advanced Theory of Statistics*, Volume 3, Charles Griffin, London, p. 48.

Lambrecht, H. (1961) 'Die Genenkarte von *Pisum* bei normaler Struktur der Chromosomen', *Agric. Hortique Genetica*, **19**, 360–401, communicated to Bennet, J. H. (Ed.) (1965) *G. Mendel Experiments in Plant Hybridization*, Oliver and Boyd, Edinburgh, p. 6.

Leyton, L. (1956) 'The relation between the growth and mineral composition of the foliage of Japanese larch (*Larix leptolepis*, Murr.)', *Plant and Soil*, **7**, 167–77. Discussed by Bliss, C. I. in *Statistics in Biology*, Volume 1, McGraw-Hill, New York, p. 308.

Lorenz, K. (1967) *On Aggression*, Methuen, London, p. 83.

MacArthur, R. H. and Wilson, E. O. (1963) 'An equilibrium theory of insular zoogeography', *Evolution*, **17**, 373–87. See also MacArthur, R. H. and Wilson, E. O. (1967) *The Theory of Island Biogeography*, Princeton University Press, Princeton.

MacFadden, B. J. (1986) 'Fossil horses from Eohippus (Hyracotherium) to Equus: scaling Cope's Law and the evolution of body size', *Paleobiology*, **12**, 355–69.

Malthus, T. R. (1970) *An Essay on the Principle of Population*, Penguin Books, Harmondsworth, England. Published in 1798 under the title *An Essay on the Principle of Population as It Affects the Future Improvement of Society . . .* This is the essay referred to here. Malthus subsequently rewrote and considerably extended the essay, which he published under the title *An Essay on the Principle of Population; or, a View of its Past and Present Effects on Human Happiness; with an Inquiry into our Prospects Respecting the Future Removal or Mitigation of the Evils which it Occasions.*

Manly, B. (1986) *Multivariate Statistical Methods: A Primer*, Chapman & Hall, London.

Margenau, H. (1950) *The Nature of Physical Reality*, McGraw-Hill, New York, p. 99. Attributed to Max Born.

May, R. M. (1976) 'Simple mathematical models with very complicated dynamics', *Nature*, **261**, 459.

May, R. M. (1981) 'Models for single populations' in *Theoretical Ecology*, Blackwells, Oxford, pp. 5–29.

Mayr, E. (1942) *Systematics in the Origin of Species*, Columbia University Press, New York. Discussed by Ruse, M. (1982) in *Darwinism Defended*, Addison Wesley, London, p. 94.

McCullogh, P. and Nelder, J. A. (1983) *Generalized Linear Modelling*, Chapman & Hall, London.

Mead, R. and Curnow, R. N. (1983) *Statistical Methods in Agriculture and Experimental Biology*, Chapman & Hall, London.

Mendel, G. (1866) 'Versuche über Pflanzenhybriden', *Proceedings of the Natural History Society of Brünn*. English translation reprinted as *Experiments in Plant Hybridization* with commentary by R. A. Fisher and biographical notes by W. Bateson in Bennet. J. H. (ed.) (1965), Oliver and Boyd, Edinburgh. See also Peters, J. A. (ed.) (1959) *Classic Papers in Genetics*, Prentice Hall, Englewood Cliffs, New Jersey, and Stern, C. and Sherwood, E. R. (eds) (1966) *The Origin of Genetics*, Freeman, San Francisco. Mendel's work is discussed by Ayala, F. J. and Kiger, J. A. Jr. (1980) in *Modern Genetics*, Benjamin/Cummings, Menlo Park, California.

Needham, J. (1972) 'Mathematics and science in China and the West', in *Sociology of Science* (ed. B. Barnes), Penguin, Harmondsworth, p. 32 citing Barbera, G. (1890) *Le Opera di Galileo' Galilei*, Volume 4, Florence.

Needham, J. (1988) 'Joseph Needham', Channel 4 Television, London, 13 August 1988.

Newsweek (1989) 14 August, 1989, p. 50.

Pacala, J. (1990) 'International doubts about an AIDS cure', *Science*, **250**, 200.

Pearson, K. and Lee, A. (1903) *Biometrika*, **2**, 357. Discussed by Snedecor, G. W. and Cochran W. G. (1987) in *Statistical Methods*, Iowa State University Press, Ames.

Pianka, E. R. (1983) *Evolutionary Ecology*, Harper and Row, London, p. 328.

Pirsig, R. M. (1980) *Zen and the Art of Motorcycle Maintenance*, William Morrow, New York, p. 238.

Powick, W. C. (1925) 'Inactivation of vitamin A by rancid fat', *Journal of Agricultural Research*, **31**, 1017–27. Sokal, R. R. and Rohlf, F. J. (1981) use the amount of lard consumed by the rats in Powick's experiment to illustrate the analysis of a two-factor experiment in *Biometry*, Freeman, New York, p. 325.

Rensberger, J. (1983) 'Trial by error', *RF Illustrated* (Rockefeller Foundation), June, p. 2–4.

Ronan, C. A. (1983) *The Cambridge Illustrated Dictionary of the World's Science*, Cambridge University Press, Cambridge, p. 429.

Ross, W. D. (Ed.) (1952) *The Works of Aristole*, Clarendon Press, Oxford. Translation of Aristole's *Physica*, W. Ogle, quoted by S. Sambursky (1974) in *Physical Thought from the Pre-Socratics to the Quantum Physicists*, Hutchinson, London, p. 65.

Roughgarden, J. (1979) *Theory of Population Genetics and Evolutionary Ecology*, Macmillan, New York, p. 21.

Ruse, M. (1982) *Darwinism Defended*, Addison-Wesley, New York, p. 150. Cites results by Fitch, W. M., 'Molecular evolutionary clocks', in *Molecular Evolution* (ed. F. J. Ayala).

Satterthwaite, F. E. (1946) *Biometrics Bulletin*, **2**, 110. Discussed by Snedecor, G. W and Cochran, W. G. (1989) *Statistical Methods*, Iowa State University Press, Ames, p. 97.

Sheppard, P. M. (1951) 'Fluctuations in the selective value of certain phenotypes in the polymorphic land snail *Cepaea nemoralis*', *Heredity*, **5**, 125.

Siegel, S. and Castellan, N. J. (1988) *Nonparametric Statistics for the Behavioral Sciences*, McGraw-Hill, New York.

Snedecor, G. W. and Cochran, W. G. (1989) *Statistical Methods*, Iowa State University Press, Ames.

Sokal, R. R and Hunter, P. E. (1955) 'A morphometric analysis of DDT resistant and non-resistant housefly strains', *Annals of the Entomological Society of America*, **48**, 499–507. Given by R. R. Sokal and F. J. Rohlf, *Introduction to Biostatistics*, Freeman, New York, p. 81. In order to simulate the results of measuring the wing lengths to one more significant figure, evenly distributed random numbers between 0.00 and 0.10 were added to each of the numbers given by Sokal and Hunter.

Sokal, R. R. and Rohlf, F. J. (1981) *Biometry*, Freeman, New York.

Sokal, R. R. and Rohlf, F. J. (1987) *Introduction to Biostatistics*, Freeman, New York.

Sprent, P. (1990) *Applied Nonparametric Statistical Methods*, Chapman & Hall, London.

Stewart, I. (1990) *Does God Play Dice? The New Mathematics of Chaos*, Penguin Books, London.

Student (1907) 'On the error of counting with a haemocytometer', *Biometrika*, **5**, 351–60. Analysed in detail by Sokal, R. R. and Rohlf, F. J. (1981) *Biometry*, Freeman, New York, p. 84.

Student (1908) 'On the probable error of a mean', *Biometrika*, **6**, 1–25.

Targett, G. A. T. (Ed.) (1991) *Malaria: Waiting for the Vaccine*, John Wiley, Chichester.

Taylor, E. F. and Wheeler, J. A. (1963) *Space-time Physics*, Freeman, San Francisco, p. 60. Quoted by Harrison, D. (1979) in 'What You Get Is What You See', *American Journal of Physics*, **47**, 576.

Tufte, E. R. (1986) *The Visual Display of Quantitative Information*, Graphics Press, Cheshire, Connecticut, p. 60. Taken from *Science Indicators 1974*, National Science Foundation, Washington, D. C., 1976, p. 15.

Utida, S. (1943) 'Studies on experimental populations of the Azuki bean weevil, *Callosobruchus chinensis* (L.). VIII. Statistical analysis of the frequency distribution of the emerging weevils on beans', *Mem. Coll. Kyoto Imperial University*, **54**, 1–22. Discussed by Sokal, R. R. and Rohlf, F. J. (1981) *Biometry*, Freeman, New York, p. 92.

von Bortkewitsch, L. (1898) *Das Gesetz der kleinen Zahlen*, Teubner, Leipzig. Discussed by Bulmer, M. G. (1979) in *Principles of Statistics*, Dover, New York, p. 92 and by Sokal, R. R. and Rohlf, F. J. (1981) in *Biometry*, Freeman, New York, p. 93.

Wallace, A. R. (1905) *My Life*, Chapman and Hall, London, Volume I, p. 232. Quoted by A. Flew (1970) in the introduction to *An Essay on the Principle of Population*, Penguin, Harmondsworth, p. 51.

Whitehead, A. N. (1928) *Introduction to Mathematics*, Thornton Butterworth, London, p. 9.

Winer, B. J. (1971) *Statistical Principles of Experimental Design*, McGraw-Hill, New York.

Index

Page numbers appearing in **bold** refer to tables

Milton Keynes UK
Ingram Content Group UK Ltd.
UKHW031149141024
449569UK00024B/937